BRITISH FIRE ENGINES
Of The
1950s & '60s

Simon Rowley

Trans-Pennine Publishing

CONTENTS

Front Cover: *This 1957 AEC Mercury/Merryweather Marquis pump (NKG 91) was supplied to Cardiff City Fire Service and operated from Roath fire station. In 1974 it passed to the County of South Glamorgan Fire Service who used the chassis and a new cab as the basis of a prime mover to carry demountable units.* John Hughes

Rear Cover Top: *A 1955 Dennis F8 pump one of 13 supplied to the Derbyshire Fire Service - this one (URB 945) served at Wirksworth.* John Shakespeare

Rear Cover Bottom: *As this book is made in Cumbria, we have to show this 1964 Bedford TK/Carmichael pump escape (BHH 415B) with the City of Carlisle Fire Service. It has since been preserved.* John Shakespeare collection

Title Page: *A pair of Bedford S type water tenders seen outside the factory fire station at the Vauxhall car plant, Luton in the mid-1950s. Further details of these appliances can be found on page 18.* Vauxhall Motors

Left: *This 1960 Bedford S pump (421 JVT) was supplied to the City of Stoke on Trent Fire Brigade at Hanley with an unusual HCB built body which contained a mixture of roll up shutters and locker doors.* Vauxhall Motors

Right: *The line up at Bury St Edmunds in the late 1960s consisted of (from the left) 1954 Dennis F12 pump escape (PBJ 338), 1954 Bedford S/Carmichael water tender (PRT 957) and 1968 Bedford TK/HCB-Angus water tender (PBJ 7F). The Fornham Road fire station was opened in 1953 and was the first new station built for the Suffolk & Ipswich Fire Service.* Simon Adamson

ISBN 0 9521070 7 4

British Cataloguing in Publication Data

A catalogue record for this book is available from the British Library

The **Nostalgia Road** Series ™
is conceived, designed and published by

Trans-Pennine Publishing Ltd.

PO Box 10

Appleby-in-Westmorland

Cumbria, CA16 6FA

Tel. 017683 51053

Fax. 017683 53558

ISDN. 017683 53684

e-mail trans.pennine@virgin.net

(A Quality Guild registered company)

Reprographics

Barnabus Design & Repro

Threemilestone, Truro

Cornwall, TR4 9AN

01872 241185

And Printed in Cumbria by

Kent Valley Colour Printers Ltd.

Shap Road Industrial Estate

Kendal, Cumbria LA9 6NZ

01539 741344

© Trans-Pennine Publishing & Simon Rowley 2000

Photographs: Simon Rowley Collection or as credited

INTRODUCTION

During the past 100 years the spectacular sight of a speeding fire engine, with its bells clanging or sirens wailing, will have turned the heads of the majority of people in the civilised world at one time or another. At the start of a new millennium it is therefore fitting to record the progress that was made during the 20th century, especially that which occurred during the 1950s and '60s, when the face of Britain's fire services changed dramatically.

As we started the 20th Century the fire engine was usually drawn by a pair of horses, and the firemen had to hang on to it for dear life as the appliance sped off on its mission of mercy. Now, at the start of the 21st Century, firefighters travel much more comfortably in 'state of the art' vehicles. Yet to the observer on the street corner, the effect is just as thrilling - a combination of man and machine with the common purpose of getting to the scene as quickly as possible and then save people from the ravages of man's oldest enemy - fire!

The development of the fire engine during this past century has been phenomenal, but that is in line with the progress made in the development of every other type of mechanical appliance.

Today's fire engine bears little resemblance to the horse-drawn steam pump. Those machines offered little protection to the firemen who put their lives on the line each time they turned out, often before they reached the scene of the fire. Yet when the internal combustion engined appliances appeared, things did not vary much for nearly another 40 years. It was not until the late-1930s that the enclosed fire tender offered any protection for the crews. Until then the men had to cling on tightly and pray that the driver did not throw them off on a sharp bend. Not only that but in the height of the winter the crews were often exposed to the worst of the elements.

Then came a new innovation, a fire tender with a limousine body that could offer covered protection for the crew. When they arrived at the incident, they were thus able to get on with the job in hand and work in dry uniforms. More importantly, not having been forced to spend all their time making sure they didn't fall off the appliance, they arrived in a calm manner ready to get on with the task at hand. It was the late-1930s before firefighters were able to take advantage of these comforts, but even so many of the old open appliances lingered on for several years as replacement programmes were restricted due to the tight financial constraints of the period.

Prior to 1941 there were some 1,526 public brigades. Some of these operated just one single fire station, whilst others had several. The largest brigade was, of course, the City of London which had about 60 stations. During the 'Blitz' every fire crew in London was stretched to capacity, and as the Luftwaffe poured down its nightly reign of terror in the form of incendiary devices, land mines and High Explosive bombs, fire crews from other towns and cities were drafted in to help Londoners cope with the vicious onslaught.

World War II placed massive pressures on the fire services, and this ultimately affected the way the brigades operated in Britain. A significant introduction that came about in 1938 (because of the impending threat of war), was the formation of the Auxiliary Fire Service, which made provision for 'reserve' firemen to train with and work alongside the regular men. Because of the difficulties caused by the Blitz, the Government then created the National Fire Service, which brought all the brigades under one umbrella. It ensured standardisation of appliances, equipment and operating procedures. Thus the AFS and all the public brigades, previously run by municipal authorities, were amalgamated into the NFS .

It is beyond the scope of this book to examine the work of the NFS during the war, but we should state that once the Blitz was over Britain never experienced devastation on such a scale again, so the organisation was never tested to its limits. Even so the V1 and V2 rocket bombs launched at Britain from occupied France frequently stretched the available resources in the South of England.

Once peace returned, the local authorities demanded their fire brigades back. But, to the dismay of many smaller councils, it was decided to give responsibility for fire cover in England and Wales to the county, county borough and city councils. Scotland was formed into 11 fire areas, and only the Glasgow Fire Brigade survived from the 'pre-nationalisation' era. Northern Ireland had its own fire authority, along with the Belfast Fire Brigade.

Thus 147 new authorities were given the task of providing fire services when the new legislation came into being on 1st April 1948. There were still some small 'one station' city brigades, whilst some rural counties had in excess of 50 stations. Yet they all had one thing in common, as their appliances were handed down from the NFS, and some were very ancient indeed. This is where our story begins!

INTO THE 'FIFTIES

Essentially this book tells the story of the first 25 years of the nationalised British fire service, beginning with the implementation of new legislation on 1st April 1948. It then covers the period up to local government reorganisation in April 1974, and forms one of the most significant periods in the history of fire-fighting.

The book compliments another **Nostalgia Road** title, *NHS Ambulances The First 25 Years* 1948-73, but the significance of the dates in this title (although appreciated by emergency service personnel), was lost on the general public at large. Therefore, we have opted in this title to concentrate on the 1950s and '60s, although the reader should be aware that the general period under consideration is 1948 to 1973. We therefore touch on the early 1970s, but the story of what happened after local government reorganisation was implemented in 1974 is, as they say, another story altogether.

Above: *Poole Fire Station was built in 1936 and this picture was taken in the early 1950s. The line up features (from left), a new Dennis F12 pump escape (EJT 678), a 1938 Leyland Tiger FT4A pump escape with limousine body (JT 9723) and a 1943 Leyland Beaver TSC/Merryweather 100ft turntable ladder pump (GXA 72). The first two appliances are preserved. Simon Adamson Collection.*

Left: *This 1936 Leyland Cub FK6 pump escape (now preserved at Taunton), was originally supplied to Yeovil UDC before being transferred to the NFS in 1941 from which it joined the Somerset County brigade in 1948. It is fitted with a 4.7 litre ohv petrol engine and 4-speed crash gearbox from which the side mounted PTO drives a Rees-Returbo 500 gallons per minute (gpm) pump. A 60 gallon water tank is enclosed in the 'Braidwood' style bodywork on which is mounted a 50ft wooden Bayley escape ladder.*

If you stood in the appliance bay of a British local authority fire station in April 1948 the first thing which you would notice about the fire engine would be its colour. However, this would not be the sparkling bright red paintwork (complemented by gleaming brass) one commonly associates with such machines from their pre-war days, but a drab all-over grey instead.

Certainly some fire tenders that came from the pre-NFS brigades had escaped the grey paint, but every appliance that had been supplied since 1941 was covered bonnet to pump in NFS grey. Now it was time to wield the red brush again and to bring a bit of colour back into the Fire Brigade. Not that colour schemes were the top of the priority list for the chief officers who led the 147 local authority fire brigades that had taken over from the NFS.

These brigades now had the responsibility of providing the public with adequate fire protection. The size of these brigades varied considerably but they all had one purpose, to save life and property from the ravages of fire.

Very soon the new brigades took stock and looked at ways of modernising their vehicles, equipment and premises. However, in the wake of an expensive war, money was very scarce. With expensive rebuilding programmes for housing, hospitals and schools, financial assistance for new fire engines was not generally made available until the early 1950s. The fire brigade fleets were therefore mainly made up of pre-war appliances and or the mass-produced vehicles supplied to the NFS, but many of these fire tenders were in urgent need of replacement.

Some enterprising brigades set about rebuilding basic wartime vehicles, and ex-military chassis with modern bodywork. As well as the brigades, fire engine builders such as Carmichael, HCB (Hampshire Car Bodies) and Miles were soon turning out rebuilt tenders on Dodge, Austin and Fordson chassis. But some of the cash-strapped brigades were forced to soldier on as best they could operating open fire tenders for many years, indeed some were still turning out to fires well into the 1960s.

Once funds started to become available the manufacturers viewed the potentially lucrative market with eager anticipation and by the mid-1950s many had seized the initiative offering modern appliances which appealed to the post-war firefighter. AEC, Austin, Bedford Commer, Dennis, and Leyland were all vying for a share of the market. There was also a growing number of bodybuilders challenging the front-runners, which included the oldest, and most famous, of them all - Merryweather & Sons of Greenwich.

Top Left: *The Vauxhall factory at Luton operated this 1939 Bedford SB escape carrier - there is no evidence of the vehicle carrying water, pump or hosereel and it is likely that it was only used to transport its crew, equipment and the 50ft. wheeled escape ladder. It is believed to have lasted into the mid-1950s.* Vauxhall Motors

Top Right: *Cheshire Fire Brigade was less than three weeks old when a rail accident occurred at Winsford on 17th April 1948. The crash, which cost 24 people their lives, was attended by fire tenders from a wide area including this 1940 Morris Commercial Merryweather pump (GMB 731) which was new to the Northwich Fire Brigade. This make of fire appliance was quite popular but was the only one of its type inherited by Cheshire. It had 'Braidwood' style bodywork named after the legendary former fire chief of Edinburgh and then London, who designed the box type coachwork on the horse-drawn manual fire engines of the mid-1800s. This style lasted over 100 years but its main drawback was that crew members had to hang on to the side-rails to avoid being flung off the speeding vehicle!* Alan Earnshaw collection

Centre Right: *An improvement in design came in the 1930s when the 'New World' style bodywork was introduced; this was much safer than the 'Braidwood' as the firemen sat in an enclosed vehicle but they were still at the mercy of the weather. An example of this vehicle was a Dennis Lancet pump (GMC 984) purchased in 1937 by Harrow UDC. It was absorbed into the Middlesex Fire Brigade fleet in 1948 and is seen here in the foreground alongside a Dennis F12 pump escape (WMX 145) at Ealing in the mid-1950s.* Charles Keevil

Bottom Right: *The late-1930s saw luxury with the first 'limousine' type of fire engine. Leyland and Dennis were soon producing fire appliances with enclosed bodies, which gained customers world wide; one such vehicle was this 1938 Leyland SFT4A pump escape, which was supplied to the Bournemouth Fire Brigade and served at Winton and Pokesdown before disposal in 1960.* Ted Hughes Collection

The domestic fire service in Britain has two main categories of fire tender - pumping appliances and special appliances; the first consists of all large self-propelled pumps, to encompass water tender (WrT), pump (P) or pump escape (PE). All other types of appliance - turntable ladder (TL), hydraulic platform (HP), emergency tender (ET), rescue tender (RT), foam tender (FoT), salvage tender (ST), control unit (CU), breathing apparatus tender (BAT), water tanker or carrier (WrC), hose layer and light strike vehicles (plus many others) - are classed as specials. Appliances operated by most airport brigades are either a major crash tender (CrT) or a rapid intervention vehicle (RIV) although the larger airport brigades have a number of ancillary fire tenders. The military brigades have domestic and airfield crash rescue appliances with their own designations.

Before World War II turntable ladders were available to British fire brigades from two German manufacturers - Magirus and Metz; obviously once hostilities threatened, these sources were no longer available and so the renowned English fire apparatus builders Merryweather & Sons of Greenwich became the only UK maker of this type of ladder.

With the Blitz causing immense damage to British towns and cities, there was an increased need for more fire-fighting appliances and so the Home Office ordered hundreds of self propelled fire tenders as well as trailer pumps, towing vehicles, escape carriers and turntable ladders. Many of the trailer pumps came from Dennis Brothers whilst Bedford, Austin and Fordson provided chassis for most of the other applications. Leyland however concentrated on building chassis for other uses but they did supply the Home Office with 29 TD7 bus chassis in 1942 (with half cabs) followed by 26 forward control TSC18 Beaver chassis in 1943; all were delivered to the NFS with 100ft Merryweather ladders. The Beaver had a Gardner 5-litre diesel engine and the appliances were equipped with 500gpm pumps mounted on the bed.

Top Left: *This 100ft Merryweather ladder was ordered by the Bournemouth Fire Brigade in 1939 and placed on an Albion chassis (FEL 823). It was fitted with a pump and gave 20 years service at the south coast town's Central Fire Station.* Ted Hughes Collection

Centre Left: *Although pictured leaving Ely, (GXA 76) passed to the Cardiff City Brigade and served at Westgate Street; it was sold to the Breconshire & Radnorshire Joint Fire Brigade in 1968 who stationed it at Llandrindod Wells. After disposal in 1971, it went into preservation although its whereabouts now are unknown.* John Hughes

Bottom Left: *The 100ft turntable ladders were far too big for many of the small towns which still required a ladder capable of tackling fires from above and so Merryweather produced a three stage manually operated 60ft ladder mounted on an Austin K4 chassis. A total of 50 were supplied and most were later equipped with front mounted Barton pumps. This 1943 example (GXN 215) was an exception - it served with the NFS at Chichester and passed to the West Sussex Fire Brigade where it lasted until 1971.* John Hughes

With deliveries of trailer pumps to AFS (and later NFS) fire stations running into the hundreds in the early 1940s, towing vehicles were at a premium. So many different types of cars, taxis, vans and small lorries were requisitioned but most of them were totally unsuitable for pulling a heavy pump. Thus came about the purpose-built auxiliary towing vehicle (ATV), the majority of which were supplied on an Austin K2 chassis powered by a 6-cylinder petrol engine. They were supplied to the NFS in large numbers and, after the war, many were converted to other fire brigade uses such as salvage tenders, foam carriers, canteen vans and emergency tenders, some lasting into the 1970s.

The *Manual of Firemanship 'A Survey of the Science of Firefighting'* issued in 1944 by the Home Office (Fire Service Department) has the following to say about towing vehicles in its Part 2 - Appliances:-
'The towing vehicle most widely used in the NFS consists of a standard medium lorry chassis, of the type used for Army vehicles, which has been fitted with a steel body built specially for Fire Service requirements. Considerable experience of the type of body required has been obtained from the construction and use of converted cars and light vans which were used as tenders by many large brigades, and it is therefore possible to produce a satisfactory design without previous experiment. The roof is specially strengthened to give protection against shell splinters and other flying fragments, and the protection afforded to the crew generally is superior to that given by the converted appliances previously used. Ample padded seating accommodation is provided and a waterproof apron is fitted by which the rear of the appliance can be completely closed to keep out rain and snow when on reinforcing duties. Ample locker space for carrying gear is provided. A 24hp 6-cyl engine with a four-speed gearbox is fitted as standard, and twin rear wheels are included in the specification.'

Top Right: *Because of the acute shortage of pumping appliances, many brigades converted ATVs into hosereel tenders complete with a small water tank and first aid hosereel. The example shown (GXH 420) served the Hampshire Fire Service at the author's local fire station at Burley in the New Forest.*

Centre Right: *Another view of the above vehicle, but this time showing the full crew including the officer in charge Sub Officer Bill Silk, as they line up for the camera during a demonstration in the local park in 1953; the appliance still towed the Dennis 500gpm trailer pump as can be seen clearly in the picture and note the canvas hose.*

Bottom Right: *First built as a pumping appliance in 1939, this Bedford K type (ANL 314) lasted until 1968 after conversion to a salvage tender by the Newcastle & Gateshead Joint Fire Service. It served at the old Swinburn Street fire station, which was eventually replaced by the station in Dryden Road in 1964. The early history of the appliance is uncertain but it is believed to have been based at Amble in the 1950s, part of the Northumberland County Fire Brigade.* Simon Adamson

Top Left: *The example shown right is typical of MDUs that found their way into the peace-time fire service at the end of the war and remained very much in the condition in which they had first been delivered. The rear facing crew shelter gave some protection to the firemen whilst the vehicle towed a large trailer pump. The Hampshire Fire Service inherited the appliance (GXM 552) and later it was rebuilt by Carmichael & Sons and served at Lymington before disposal in 1966.*

Centre Left: *The Carmichael rebuild is illustrated by the centre picture which shows another Hampshire machine (GXO 564) that came to Burley in 1954 replacing the Austin K2 ATV shown on the previous page. Constructing bodywork and fire engineering on wartime chassis was a cost-effective way of providing water tenders when money was scarce and there was a limited supply of new vehicles. As well as Worcester-based Carmichael, Alfred Miles of Cheltenham undertook rebuilds on Dodge and other chassis for many fire brigades.*

Bottom Left: *Not only did the Hampshire Fire Service contract out the work on its fire engine rebuilds, but they also rebuilt a quite considerable number of Dodge chassis in their workshops at Winchester. The first was built as an 'A' type water tender (with no plumbed-in main pump) in 1949, followed by 'B' types (with power take off to an on-board pump). A front mounted Barton pump was fitted to this 1953 rebuild (GXM 637) which was allocated to New Milton.*

The British Government was conscious about the immense damage, which was being inflicted by enemy action, and often affected water supplies. It was therefore decided that a fleet of lorries fitted with dams should be supplied to the NFS. Many of these vehicles were built on Dodge 82A chassis where the '82' referred to its weight in cwt (i.e. about 5 tons). The chassis were built in London at Kew to a 1939 American design and were fitted with 6 cylinder Dodge side valve petrol engines and a 4-speed crash gearbox. On the flatbed was mounted a 500 or 1000 gallon dam and a Worthington Simpson 250gpm pump powered by a Ford electric start engine.

The 1944 *Manual of Firemanship* advises that 'the dam lorry is used as an emergency method of bringing water to a fire where only relatively small quantities are required, or where a hose relay would be impracticable.' This vehicle was then the forerunner of the water tender which, for the last 50 years, has been the workhorse of the British Fire Service.

With money so scarce, the rebuilds of wartime vehicles were very useful as a temporary cost effective measure. However, the chassis and bodywork manufacturers wanted to produce fire appliances which would both suit the needs of the modern fireman but at the same time did not give the county council chief accountant a headache when the bill dropped on his desk.

The design teams at AEC and Merryweather got together and between them developed a purpose-built fire appliance using a shortened version of the AEC Regent III bus chassis. It had a cab built by Park Royal with the bodywork and fire engineering provided by Merryweather. By 1951 their advertisement in the July edition of *Fire Protection and Accident Prevention Review* proclaimed that 'its many outstanding features have made it the choice of fire brigades throughout the country. One of these features - its rear stabiliser - effectively reduces excessive roll' and steadies the machine when cornering at high speed.'

Dennis, meanwhile were very keen to regain their former stronghold in this specialist field and so they introduced a new lineage in 1946 known as the 'F' series, which was to continue in various derivatives until the 1970s. The F1 and F2 models were built to an open design which, due to overseas sales, lasted in production until the 1960s. However, as the designers quickly made improvements, by the end of the 1940s the first 'modern' Dennis appeared in the form of the F7. Powered by a Rolls Royce B80 petrol engine coupled to a 4-speed manual gearbox, the F7 was a powerful machine but brigades found that its wheelbase was too long and it was unstable on corners. Soon it was followed by the highly acclaimed F12 with a 150-inch wheelbase; it carried 100 gallons of water and had a Dennis No 3 pump. This delivered 1,000gpm if it was rear-mounted or 900gpm side mounted. Most carried a 50ft wooden wheeled escape ladder which, on the machines fitted with a rear mounted pump, had to be slipped off before the pump operator could get to work!

Also anxious to gain a strong foothold was the Rootes Group who introduced their Commer FC 21A fire tender chassis in 1950. It had a 6-cylinder petrol engine and a 50gpm PTO driven first-aid pump fed from the 400-gallon water tank. Its main pump was a separate unit thus making it an 'A' type water tender although most were converted in the mid-1950s to 'B' types with a PTO driven in-built major pump.

Top Right: *Part of an order of 60 supplied to the London Fire Brigade, this AEC Regent III/Merryweather pump escape (PUC 165) is fitted with a 1,000gpm pump mounted midships. This was one of the last batch delivered in 1955 and it served at Canon Street and then Burdett Road (Bow) and lasted until 1973.* Simon Adamson

Centre Right: *The majority of Dennis F12s carried 50ft wheeled escape ladders but by the time the picture had been taken in the late 1960s, this appliance (FT 8008) was carrying a 35ft triple extension ladder thus classifying it as a 'pump'; it was operated by the single station Tynemouth Fire Brigade.* Simon Adamson

Bottom Right: *Still in the North East but this time carrying bodywork by James Whitson of West Drayton, Middlesex, we show a Commer FC 21A (DJR 940), which was originally built for the Northumberland County Fire Brigade as an 'A' type water tender and was stationed at Alnwick. It was later fitted with an in-built pump, thus converting it to a 'B' type WrT.* Simon Adamson

Below: *Once peace returned, the German turntable ladder manufacturers of Metz and Magirus hoped for buyers amongst the British fire brigades but it was not until the early 1950s that they started competing against Merryweather. Magirus ladders were marketed by John Morris of Salford using the Leyland Beaver chassis. This was followed David Haydon of Birmingham using Bedford or Commer chassis. Metz, on the other hand chose Dennis Brothers of Guildford to mount their ladders and the 'F' series was expanded to accommodate ladders ranging from 100ft to 125ft. The City of Nottingham used a Dennis F21 (YAU 999) for its 100ft Metz ladder delivered in 1955 and allocated to the Central Station. It was transferred to Dunkirk in the Nottinghamshire Fire Service and is now preserved. It is worth noting that pair of Dutch-built Geesink 107ft turntable ladders were purchased in 1951 by the West Riding Fire Brigade in Yorkshire; but as far as we know they were the only ones to be bought by a British brigade.*

Above: *Merryweather & Sons were anxious to find a suitable chassis for their 100ft 4-section turntable ladder, and in 1947 turned to AEC who provided their Regal coach chassis powered by a 9.5 litre Meadows petrol engine. Following an order for 10 appliances placed by the Home Office, the first one appeared in 1949 for the Surrey Fire Brigade. Eight of the TLs were fitted with pumps and some of the appliances later had their petrol engines replaced by diesel versions. The one illustrated (LYB 388) was delivered to the Somerset Fire Brigade in 1950 and went on the run at Weston-Super-Mare where it lasted until 1974. It was sold to a cleaning firm in Worcestershire who repainted it yellow but it has now been fully restored to its former condition.* John Hughes

It seemed that just about everybody wanted to build fire engine bodies on Commer chassis. The main players were naturally Carmichael & Sons of Worcester, Hampshire Car Bodies of Totton near Southampton and Alfred Miles of Cheltenham but anxious to display their expertise were many other coach-builders, some of whom had no previous experience in the rather complicated sphere of fire engineering.

The advertisement that appeared in the June 1953 edition of *Fire Protection Review* (promoting the Commer fire appliance chassis), illustrated six appliances with five different bodywork manufacturers. As well as a Miles water tender and pump escape, and water tenders from Carmichael and HCB, there were also water tenders from Jennings of Sandbach and Cuerden of Blackburn.

The chassis embodied a 109 bhp 'under-floor' engine with long life full-length chrome finished cylinder bores. Above the banner 'Proved in Service the World Over' the text explained that, working in close collaboration with fire authorities, Commer had evolved a chassis incorporating all the essential requirements for tenders, pumps & pump escapes. It added 'inherent features of the power unit are quality, reliability and performance, and the chassis - fully proved in service - conforms to the essential requirements of the appropriate Home Office specification and is backed up by a nation-wide service organisation.'

Also advertising bodies for the Commer chassis were the firms of John Kerr & Co from Liverpool ('Kerr-fully built') and Windovers of Hendon, London. But there were also a number of manufacturers who built Commer bodies for their local brigades; these included Vincent (Berkshire), Cooper (Suffolk), Button (Monmouthshire) and two in Kent, Kennex and Martin Walter.

Top Right: *During 1951 the Berkshire & Reading Fire Brigade placed an order with a local bodybuilder, Vincents, who they commissioned to build bodies for a Leyland Comet and three Commer 21A appliances. One of the Commers (GRX 34) was built as a PE and it is seen here in its later form as a pump ladder. It served its entire career at Windsor.* Simon Adamson

Centre Right: *The working end of the same appliance clearly showing the conversion made in 1967 to change from carrying a 50ft wheeled escape to a 45ft triple extension ladder. The stowage for the 4inch suction hose is unusual - normally these are housed in tunnels that run the length of the appliance body.* Simon Adamson

Bottom Right: *The builders Alfred Miles boasted that their appliances were 'the outcome of unique experience in the design and construction of modern aircraft.' By saving well over a ton (and in some cases over 30cwt) of unnecessary weight over traditional methods of construction, the power/weight ratio of the Commer-Miles all alloy vehicle is increased by as much as 25%'. The 1951 advert claimed that their appliances had been chosen by over 20 authorities and these included Wiltshire whose 1953 WrT (JHR 172) served at Tisbury and is seen running as a spare at Salisbury.* Dennis Hill

Top Left: *This interesting 'fireground' picture from Somerset in the early 1960s shows a 1954 Miles bodied Commer 45A WrT (RYA 601), allocated to Winscombe, with its lockers open, showing the canvas hose, and the first-aid hosereel tubing pulled out. On the roof is a 35ft double extension Bayley ladder and also seen are the suction hoses - 4-inch for the main pump and 3-inch for the light portable pump stowed under the cover on the rear offside. The onboard Dennis pump possibly dates from a wartime trailer pump. The vehicle on the right is a 1939 Bedford towing vehicle with a Coventry Climax light trailer pump. This combination served the small village of Banwell, which provided one of two volunteer fire units in Somerset but were closed down soon after they were absorbed into the County of Avon Fire Brigade in 1974. In* The Story of the Banwell Fire Brigade, *Wally and Roy Rice recall that the Bedford 'was reckoned to previously have been a coal lorry ...... it had boxes built behind the side-boards of the lorry-bed leaving a gangway up the centre for personnel to stand. For those of us who had the privilege (or otherwise) of trying to keep balance and put on boots, tunic, etc, while 'the Bedford' manoeuvred at speed around Banwell's corners on the way to a fire - it is an experience we will never forget!'* Ian Scott

Centre Left: *The Austin K4 Loadstar never gained popularity amongst the British fire service with the exception of Cornwall, which actually had a fleet of 30. These appliances were built at the Home Office workshops in Swindon at a time when Cornwall County Council was experiencing serious financial problems. Advertisements from 1951 and 1954 show John Kerr and Carmichael offering Loadstars but only a handful were built although horsebox builder Lambourn is believed to have built seven for Oxfordshire and a few went to Northamptonshire. The vehicle illustrated is a 1953 pump escape (NRL 361) which served at Camborne in Cornwall. It came out of service in the late 1970s and, after a period of preservation, was last seen with New Age travellers still carrying its 1938 Bayley heavyweight escape ladder.* Gary Chapman

Bottom Left: *The wartime Austin K6 chassis was chosen as the basis for the first purpose-built airfield crash tender produced since World War II, and the resulting appliances carried 600 gallons of water and 60 gallons of foam. The Ministry of Transport and Civil Aviation ordered a fleet of Pyrene-built 6x4 pump-foam tenders which were distributed to civil airports and research and development establishments throughout the UK. This one (HXA 893) had a 1953 body on a 1941 chassis and when seen in the 1960s was at the MTCA training school at Pengam Moors, Cardiff, having previously served at Royal Aircraft Establishment in Farnborough (see page 16).* John Hughes

Top Right: *This splendid view shows an early AEC Regent III pump escape delivered to the Glamorgan Fire Service in 1950. The appliance has a Park Royal cab and has been built by Merryweather & Sons at their factory at Greenwich in south east London. The appliance is fitted with a side-mounted Merryweather pump (either 500gpm or 1,000gpm) and it carries a 50ft Merryweather steel wheeled escape ladder. In line with buses and lorries using a similar chassis, the Regent was powered by an AEC 9.6 litre direct injection diesel engine thus making it the first production fire appliance to be diesel powered. It was fitted with an AEC D124 four speed crash gearbox. A total of 114 pumping appliances and one emergency tender were built for UK fire brigades using the shortened Regent chassis from 1950 until 1956 and there were also five turntable ladders using the normal Regent bus chassis. This particular vehicle (KTG 617) was stationed at Pontypridd for most of its service life but when this picture was taken at a fire in Rhymney in the 1960s, it was operating as a reserve appliance based at Bargoed.* John Hughes

Top Left: *Factories (and other industrial sites) which had high fire risks have, since the beginning of organised fire fighting, operated their own fire brigades. Although many of them used (and still use) second-hand fire tenders purchased from local authority brigades, several firms ordered new vehicles. The Pyrene Company supplied this Bedford SHC pump/foam tender (NNX 872) to the General Electric Company at their Willians Road Works in Rugby in 1953. The other GEC plant at Rugby operated a 1952 Dennis F2 PE.* Vauxhall Motors

Bottom Left: *A splendid line up of fire appliances operated by the MTCA at the Royal Aircraft Establishment at Farnborough in the mid-1950s. The vehicles are (from left), a general purpose lorry, a 1956 Thornycroft Nubian 6x6/Sun Engineering foam and CO2 tender (SXT 112), a Bedford QL water bowser, an Austin K6/Pyrene pump/foam tender (HXA 893), an Austin K2 ATV (GLR 433) and an Austin K6/Sun Engineering pump/CO2 tender (JUV 657).*

Top Right: *The first purpose-built airfield crash tender built for the RAF Fire Service after World War II was delivered in 1951 on a Thornycroft Nubian TF 4x4 chassis by Sun Engineering of Kingston-upon-Thames, although it is believed to have been built by James Whitson. It was designated as a Mark 5 and approximately 80 were built until 1955 when replaced by the Pyrene-built Mark 5A. Powered by an 8-cylinder Rolls Royce B80 engine, the manufacturers claimed the vehicle could deliver 2,500gpm of foam or 200gpm of water or both at the same time. It carried 400 gallons of water and 60 gallons of foam, seen here is 45 AF 53.* Simon Adamson Collection

Centre Right: *It was more than 10 years after World War II that civil airports received new crash tenders ordered by the MTCA. These were Thornycroft Nubian 6x6 appliances supplied by Sun Engineering. There were three types - major foam tender (800 galls water 100 galls foam), water tender (960 water 35 foam) and foam/CO2 tender (550 water 100 foam plus 600lb CO2). The latter type could deliver 7,500 gpm of foam or 100lbs per minute of CO2. The Ministry of Aviation Fire Service had its headquarters in The Strand, London and was responsible for fire fighting at 20 main airports as well as research and development establishments and Royal Ordnance Factories. Bournemouth (Hurn) Airport operated a foam/CO2 tender (VXN 866) which later went to Dan Air at Lasham Airfield in Hampshire and is now preserved at the Brooklands Museum in Surrey.*

Bottom Right: *Although dozens of Thornycrofts were supplied for airfield fire fighting, the Basingstoke-built vehicles found little favour with local authority fire brigades with only about 20 being purchased. Carmichael built a batch of Nubian 4x4 WrTs for Staffordshire and a number were supplied to Glamorgan. They carried 500 gallons of water and the major pump delivered up to 900gpm. This one (GFL 507) was new to the Soke of Peterborough Fire Service and served at Dogsthorpe but is seen here as part of the Huntingdon & Peterborough County Fire Service fleet.* Simon Adamson

BEDFORD

When statisticians add up the totals of different fire engine chassis supplied to British fire brigades during the 25 years after World War II, there is no doubt that the manufacturer who supplied the most was Bedford. It is therefore appropriate to look at the role that this manufacturer played in the development of the British fire appliance.

Bedford was part of the General Motors' empire and was basically the commercial vehicle wing of Vauxhall Motors, and shared a vast site at Luton in Bedfordshire from where the first Bedford truck was launched in 1931. During the war years the production of Vauxhall cars was cut back at the order of the Government, and most of the factory was turned over to truck and tank production. Most of the trucks went to the armed forces, and around a quarter of a million were built in the six years between 1939 and 1945. On the tank side, the works designed and built no less than 5,640 Churchill tanks, each of which weighed 38-tons. At the end of the war, and in a desperate need to earn foreign currency, the Government ordered the company to export the bulk of its production. and 10,000 trucks were sent overseas in the period up to 31st October 1946. Due to this massive pressure to export, there were simply too few new chassis for the fire services.

However there was a substantial re-building programme for ex-WD wartime Bedford (mainly on the QL and the O series) chassis. This work was undertaken at various locations, including the Home Office Workshops at Wakefield in the West Riding of Yorkshire. Meanwhile the Luton plant also supplied as many new chassis as it was allowed under the restrictions of post-war Britain, but it was to be the late-1940s before things got easier..

As funds for new fire tenders started to become available after the war, rural brigades throughout the UK were anxious to find a cost effective fire appliance which suited their needs for a water tender. This would have to carry 400 gallons of water, a built in major pump, a full complement of fire fighting equipment, including a 35ft double extension ladder, plus of course a crew of seven firemen. In the first instance the Rootes Group's Commer had seemed the obvious choice for the new orders being placed after nationalisation, but when Bedford announced its 7-ton S type chassis in 1950, chief fire officers were very interested in what the designers had to offer and they were not disappointed. At first the Bedford coach chassis variant, the SB, was adopted. This was then modified for use as a fire tender, with the 17ft 2in wheelbase shortened to 12ft 6in by Baico.

Another Birmingham firm, Prestage, did the fire engineering. Wilsdon of Solihull and Oldland of Bristol bodied a number of examples, mostly as pump escapes. By 1952 the SH was the preferred chassis for water tenders and the lengthened SL provided the basis for Morris-Magirus 100ft turntable ladders. Meanwhile the 4-wheel-drive R type, although finding only a small niche in the local authority market, found favour with the Home Office who ordered hundreds for the AFS as emergency pumps in addition to the S type 'Green Goddesses'. The small A type appealed to several industrial and a few county fire brigades whilst, from the late 1950s, the J type in its various forms was a very popular choice with brigades, large and small, well into the 1960s. A new factory at Dunstable was opened in 1955 and by the end of the decade the company produced about a quarter of the commercial vehicles sold in the UK as well as being the largest exporter of truck chassis in Britain. Although many of the lorries and vans sold by Bedford during this time were powered by diesel engines, the vast majority of fire tenders were fitted with a Bedford 6-cylinder 214 or 300 cubic inch petrol engine. Following the R, A, D, C and J types, Bedford introduced their renowned TK series, which was probably the most popular fire engine chassis ever produced.

Top Left: *Vauxhall Motors operated its own works fire brigade which, during the mid 1950s, used a pair of Bedford S type water tenders including this 1953 Miles bodied example (KTM 452). The other one (LMJ 614), which is seen on the title page, was one of a few fire tenders to receive a body from Kent coach-builders Martin Walter. Also in the fleet was a Bedford CA van (PBM 678).* Vauxhall Motors

Top Right: *Miles also built the bodies for a large fleet of domestic water tenders supplied to the RAF Fire Service using the Bedford SHZG5 chassis. The all red vehicles carried much the same type of equipment as the civilian water tenders but they were fitted with 500-gallon (as opposed to 400-gallon) water tanks. This 1958 example (22 AG 78) was the third production model of 80 built; it saw service with the RAF in Episkopi in Cyprus and was later transferred to the Army Fire Service before withdrawal in 1975.* Vauxhall Motors

Centre Right: *This lengthened Bedford S chassis (the SL) was the first of its type on which a Magirus 100ft all-steel turntable ladder was fitted; John Morris & Sons of Manchester supplied the bodywork and the engineering was carried out by Prestage. The appliance (VTJ 416) was delivered to the Lancashire County Fire Service in 1955 who stationed it at Leigh before it was sold, in 1967, to the Isle of Wight Fire Brigade for service at Newport.* Vauxhall Motors

Bottom Right: *This Bedford S type built by Hampshire Car Bodies as a water tender shows it clean lines in a side view taken outside Ferndown fire station in Dorset where the appliance (GFX 603) was stationed from 1954 until 1966. All the crew-members, including the driver and officer in charge, had to make their entry to, and exit from, the cab through the jack-knife doors.* Vernon Sauer

Above: *This pre-delivery photograph shows the Bedford R type 4x4 water tender to be shown at the Chief Fire Officers' Conference at Southport on 16th January 1957. The Wilsdon bodied appliance (SCE 118) was supplied to the Cambridgeshire Fire Service and served at Soham. Although hundreds were made for the AFS, the R series was not a popular choice for local authority brigades with only 16 pumping appliances and 11 specials supplied.* Vauxhall Motors

Left: *The later style of the Bedford S type radiator grill, introduced in 1957, is shown in this view of a 1960 HCB-built water tender (GJS 511) which served with the Northern Area Fire Brigade in Scotland at Kyle of Lochalsh. Fire appliances also used the C type chassis and from an outwards appearance it was impossible to tell the difference between the two types; indeed for years the author believed his own former UKAEA (Harwell) pump/CO2 tender (PJT 508) was an S type but when he checked the identification plate during the course of writing this book he found it was a C5Z3!* Alistair MacDonald

Top right: *The Oldland Motor Body Builders Ltd. of Bristol supplied their local brigade with a number of pump escapes that used the shortened SB bus chassis. These carried 100 gallons of water and were fitted with a Dennis No 3 pump. The PE pictured (150 BHW) dates from 1958 and served at Brislington until it was written off in a serious accident in 1968. About four dozen SBs were built as fire appliances by various bodybuilders with the Birmingham Fire & Ambulance Service receiving the most - 27 with Wilsdon bodies.* Charles Keevil

Middle right: *The Suffolk & Ipswich Fire Service purchased several water tenders using the Bedford TA range 4-ton chassis, which were bodied by Carmichael. This example (TRT 623), which served at Sudbury, dates from 1956. Fire appliances of this type were converted to semi-forward control by Nevilles of Retford.* Simon Adamson

Bottom right: *Alfred Miles used a standard body style for the Dennis F8 and Bedford A and J types. This Bedford A4S/Miles water tender (SUK 999) was supplied to the Goodyear works fire brigade at their Wolverhampton factory in 1957 and it lasted into the 1980s.* John Shakespeare

From a difficult relationship with the body-builders and fire brigades in 1946-1948, Bedford courted both parties in advance of their introduction of the S range, and the tactic worked well!

Whilst talking of bodybuilders it should be said that, whilst this book does not concentrate on the history of the individual firms, we can not escape the fact that there are quite a few interesting stories to be told here. As the **Nostalgia Road** series is always looking for new titles, this is an area to which we may return. For the moment we might briefly consider just a few of those firms whose products we illustrate here.

For example Alfred Miles Limited was a popular choice of bodybuilder on the Bedford chassis throughout the 1950s but one manufacturer that had a very close affiliation was Hampshire Car Bodies of Totton near Southampton. The two companies worked together on producing a number of highly popular variations of the J series chassis. Also in Hampshire, the county's own fire brigade was one of very few fire authorities producing bodies for its own fire appliances. This ran for seven years, from 1955 until 1961, during this time it standardised on the S type for its pump escapes (and later its water tenders) building a total of 23 in its Winchester workshops.

An interesting experiment was covered in the June 1955 edition of *Fire* magazine where the Oxfordshire Fire Service commissioned an appliance designed by their chief 'to meet the special requirements of rural areas.' Based on a Bedford A type 4-ton chassis and fitted with a Dennis 500gpm pump, it was built by Lambourn Engineering. The article said that 'its compactness - it had an overall length of less than 20ft and a turning circle of 41ft - made it highly manoeuvrable.' However one wonders if its 200-gallon tank was really adequate for an appliance on a rural fire station and certainly very few of these were built either for Oxfordshire or other predominately rural brigades.

THE AUXILIARY FIRE SERVICE

In the first instance it was the Air Raid Precautions Act of 1937 that gave provision for a volunteer force to supplement the regular brigades in the event of war and this brought about the formation of the Auxiliary Fire Service. The AFS was brought under the umbrella of the newly formed National Fire Service in August 1941, with the intention that all brigades in the UK operated under the same unified command. This situation remained until April 1948 when brigades were returned to local authority control although at that stage the AFS was not reformed.

However international unrest was worsening and the fragile peace was threatened by a world-wide nuclear holocaust. Thus the Civil Defence Act was passed in November 1948, and this authorised the reconstituting of the AFS and the Civil Defence. Under the auspices of the local authority brigades, AFS recruitment and training started in late 1949. In the first instance it mainly used wartime fire appliances such as heavy units and ATVs, along with trailer pumps.

Above: *A Bedford RLHZ 4x4 'Green Goddess' emergency pump lifts water from a portable dam during an AFS exercise at the Fire Service Research Centre in 1958. This is one of the water relay booster points placed every half a mile in order to keep up the pressure, using an open circuit where the water was pumped into a collapsible dam and sent on to the next one.* Vauxhall Motors

In the event of a nuclear attack it was accepted that the target areas would suffer total devastation, and this would in turn overwhelm the local emergency services. It was planned to use 'mobile columns' that were located away from likely targets; these would be self-contained and would respond to the affected areas.

As in World War II, the fire service would again be nationalised by means of legislation to cope with the 'national emergency'. The wartime fire vehicles were not suitable, so new emergency fire appliances and equipment were designed and these were issued to the local authorities or were stockpiled at various locations throughout the UK.

The new emergency pumps were based on the Bedford SHZ 4x2 chassis, but they were soon followed by a 4x4 version on the Bedford RLHZ chassis. Both types were equipped with Sigmund FN4 900 gpm pumps with four deliveries; the 4x2 versions had 400-gallon tanks as opposed to the 4x4s which carried 300 gallons. Equipment, similar to that carried by local authority water tenders, was stowed in the large lockers with lift-up doors. Thousands of the AFS appliances (nicknamed Green Goddesses) were produced by a variety of bodybuilders all to an identical design, so it was impossible to tell which builder had built which appliance especially as only a few fitted makers' plates. What is known is that ten bodybuilders were responsible for the majority of the appliances; these were Harrington, Hoskins, Papworth Industries, Park Royal, Plaxton, Strachan, Weymann, Whitson, Willowbrook and Windover. It is interesting that many of these were bus or coach builders and the major fire engine builders (Carmichael, Miles and HCB) were not involved.

The mobile column also needed a variety of other vehicles to ensure that it was a self contained unit so hose layers, pipe carriers, foam carriers, control units, canteen vans, recovery vehicles, communication vans, transportable water units and general purpose lorries were among the types provided. As well as Bedford S and R types, the Commer Q4 was used for various applications and more than 250 of these were supplied.

In addition to the large vehicles, there was also a need for small 4x4s to carry out a variety of tasks including command, communications, general purpose and reconnaissance and so Land Rovers and Austin Gypsies were supplied along with BSA, Matchless and Triumph motorcycles.

There is no doubt that the original recruitment targets were never reached but in the event, as tensions eased and threat of war diminished, the mobile columns were never tested 'in anger' although throughout the 1950s and early 1960s exercises were held at regular intervals. There were also times when AFS personnel were requested to help out in peacetime emergencies such as the sinking of the MV *Torrey Canyon* bulk oil tanker, which caused widespread oil pollution to the Cornish Coast in March 1967.

During the 1960s, America and Russia, the main perpetrators of the 'Cold War', made considerable progress to ensure world peace and so the British Government of the day decided that there was no further need for the AFS and ruled that it should be stood down. Thus the AFS was disbanded in 1968. Some of these appliances were purchased by local authorities or works fire brigades, but the majority went into store. Many of these have since been sold off but a proportion of 4x4 Bedford EPs are still retained for emergency use and in recent years they have been seen in action during flooding, at heathland fires or been used by military personnel during fire service disputes.

Above: *The Commer 3-ton Q4 had various AFS applications including the well known 'bikini' unit in which it transported portable pumps and RFD inflatable rafts; this picture shows a hose layer (SXF 530) which was used by the AFS at Bury St Edmunds, Suffolk.* Simon Adamson

Below: *Bedford 4x4 EPs were also used by the Army Fire Service - they were very similar to the AFS examples but had some minor differences. This one (18 CE 65) was based at Bordon Army Fire Station in Hampshire. It was sold in 1974.* Simon Adamson

COMMER KARRIER

In the post-war period the Rootes Group, looked at ways of expanding their commercial vehicle market. The company had long enjoyed a good toehold in the municipal market, mainly with regard to refuse collection vehicles and park's department lorries, but they also had some experience with supplying fire engines. However, they knew there was a bigger market in the mid-size range where Dennis and Bedford were getting a full share of the sales. Rootes viewed this as potentially dangerous to their sales of chassis for other municipal applications, fearing that councils might well standardise on a single make. Accordingly they launched their challenge with the Commer-Karrier Gamecock in 1956, after close consultation with Carmichael. It had a 3- 4-ton payload and a 9ft 7in wheelbase making it ideal for rural as well as city brigades. Despite its compact size, the appliance was still able to carry 400 gallons of water. It was powered by a Commer 6-cylinder 4752cc under-floor petrol engine. As was the usual (with pumping appliances being built at that time), many Gamecocks were equipped with second-hand pumps recovered from wartime trailer pumps.

Most of the vehicles were built with a pair of large side lockers, although the pump escapes for the Northern Area incorporated three lockers a side. The lockers were fitted with roller shutters, which were becoming the standard practice replacing hinged doors. A feature of some Gamecock water tenders was the large roller shutter, which encased the pump bay at the rear of the vehicle. Many were also panelled in embossed aluminium to cut out the cost of painting the vehicle with the exception of the front end and cab roof. As opposed to many of the Commers, the Karrier models incorporated cab doors for the driver and officer in charge so they did not have to clamber out through the jack-knife doors after the remainder of the crew had climbed out. The Gamecock found orders with several brigades and total sales in the UK reached about 100 but there were plenty of overseas buyers notably in New Zealand and South Africa.

Top Left: *Only seven Gamecocks were purpose built as special appliances although six Miles-built hose carriers were later converted into emergency tenders by Northumberland and a couple of WrTs were turned into foam tenders. Middlesex commissioned Waldergraves to build bodies on several special appliances although the author knows nothing about this firm (information would be appreciated). Middlesex also fitted out this 1958 Gamecock in their extensive workshops at Ruislip. Salvage tender (20 LMX) was based at Feltham and Ealing before passing to the London Fire Brigade in 1965. Charles Keevil*

Centre Left: *An interesting comparison of front ends at the Durham County Fire Brigade HQ station at Framwellgate Moor; here a 1957 Karrier Gamecock 72A/Carmichael WrT (YPT 696) stands alongside a 1961 Bedford TJ4/HCB WrE (204 JUP). Simon Adamson*

Bottom Left: *This all-white 1957 Karrier Gamecock 72A (RWO 698) was built for the Monmouthshire Fire Brigade by T. & G. Button as a towing vehicle for a trailer pump. It was later converted into an emergency tender for Pontypool and lasted until 1974. John Hughes*

Above: *'Metal or Wood?' asked the headline in the Alfred Miles advertisement in the June 1953 edition of* Fire Protection Review. *Then it went on to claim 'There is NO comparison because metal is ideal for carrying fluctuating and concentrated stresses; metal is impervious to temperature changes; metal is of reliable consistent section (no grain, knots or other imperfections); metal will not rot or shrink; joints do not become weak or loose; a metal section can be made as strong as an equivalent wooden section four times its weight; in the event of an accident, metal retains its strength, however badly deformed and localised the area of damage. Wood will transmit an accident impact, losing strength and splintering.' Accompanying the text was a picture of a Bedford S type water tender escape with a lightweight all-aluminium body similar to the example above. However the 1958 West Sussex WrE (5326 BP), which served at Storrington, has the later grill and also an unpainted body. The idea of embossed aluminium panelling started in the mid-1950s with an aim to cut the cost of painting and also because it was more durable than painted body panels. It was pioneered by the Kent Fire Brigade and because it was so successful, other brigades followed suit.* John Hughes

Above: *The County of Hereford Fire Brigade were one of the first buyers of the Dennis F12 pump escape, their example (HCJ 839) being placed on the run at Hereford in 1951. The wooden Bayley 50ft wheeled escape ladder is clearly seen as are the three deliveries of the Dennis No 3 pump. It is now preserved in South Wales.* John Hughes

Left: *The compact Karrier Gamecock bodied as a Carmichael pump escape is well illustrated by this 1956 example (KST 127), which served the Northern Area Fire Brigade in Scotland and was stationed at Inverness. Its bright red livery is so typical of British fire appliances but fire engines were not always red as we will now show. In the 1960s experiments carried out by the Lanchester College of Technology in Coventry (in conjunction with the Home Office), led to at least three brigades painting their appliances yellow.* Simon Adamson

CHEMICAL INCIDENT UNIT

Above: *Coventry took the lead followed by West Sussex and Newport (Monmouthshire). Using a specially prepared ICI paint 'Fire Brigade Yellow', they felt that a bright yellow livery would make the fire tenders conspicuous at the scene of an incident and thus lessen the danger of it being involved in a collision. Illustrated is Newport County Borough Fire Brigade 1972 Dennis F48 emergency tender (WDW 202K) which served at Malpas. The picture was taken soon after the brigade was absorbed into the Gwent Fire Brigade in 1974. Although it was later converted into a decontamination unit and then to a chemical incident unit, it retained its yellow livery until disposal in 1986.* Barrie Lowe

Below: *Another variation was the white painted fleet of the Bedfordshire Fire Service shown by this 1971 Ford D600 water tender ladder (TNM 449K) which had Merryweather bodywork and was stationed at Kempston.* Martin West Collection

In a time span of 25 years, the period between nationalisation in 1948 and local government reorganisation in 1974, the face of Britain's fire service changed dramatically, a fact well demonstrated by these two pictures:-

Above: *The fire-fighting hydraulic platform was first introduced into the UK in 1963 and for a time the Simon Snorkel 50ft version was popular. The twin booms were housed on the top of a conventional pumping appliance, as can be seen in this view of an ERF 84PF with HCB-Angus bodywork (SVE 668K). It was new to Cambridgeshire in 1972, but was soon sold to the Monaghan Fire Brigade in the Irish Republic (902 BI).* Martin West Collection

Below: *Typical of a post-war body built by the Home Office Workshops at Wakefield onto wartime chassis is this Austin K4 (GJJ 833). The appliance was originally supplied as an escape carrier to the NFS in 1941 and was inherited by the South Western Area Fire Brigade (Scotland). Equipped with a front mounted Barton pump, it carries a pair of Ajax ladders - a 35ft extension and 50ft wheeled escape. After service in Stranraer, it was saved for preservation and totally restored by Dave Smith of Great Driffield and is now at Eden Camp, Malton, North Yorkshire.*

FOUR-WHEEL DRIVE VERSATILITY

Soon after the war it became apparent to manufacturers that a small 4x4 vehicle would be of great benefit for many fire brigades, because it could ferry water and a crew of four to inaccessible locations. In 1948 Land Rover produced a fire tender based on its 80in pre-production model, and this was followed in the early 1950s by a fire engine conversion to the Series 1 86in. Then, in the late-1950s, Carmichael & Sons were given the sole right to manufacture Land Rover fire tenders, an arrangement that lasted for three years after which other manufacturers joined the fray.

The 86in fire tender carried a rear mounted twin delivery self-priming Pegson 200 gpm pump and a first aid hosereel was fitted in the rear body above a 40-gallon water tank. Equipment lockers were fixed above the rear wheel-arches and suction hoses were carried on the bonnet but this was not ideal as the bent hoses often cracked. A ladder was carried on a gantry, which was fixed to the front bumper and rear body, and this was later adapted to carry the suction pipes. Although popular with works fire brigades, the 86in found only a few buyers with local authorities who preferred the 107in format. Dispensing with the main pump and, using just the first-aid pump and hosereel, it was ideal for tackling heath, moorland and forest fires. The Series II fire engine versions in both short (88in) and long (109in) wheelbase formats found many buyers at home and abroad.

BMC also wanted a piece of the action and they quickly offered their Austin Champ in fire engine form. Based on the civilian WN3 version, and powered by a 2660cc 75hp 4-cylinder Austin A90 petrol engine. The London-based Fire Armour company produced some as foam-CO_2 crash tenders as part of the Firefly range. The Champ found few takers but in 1958 Austin's Longbridge plant had a new product to challenge the Land Rover, namely the Gipsy. It was produced over 10 years as series 1, 2 and 4 and in short or long wheelbase form, and there were fire engines of each variation. Powered by a BMC 2.2 litre A70 petrol engine, the Gipsy had a front mounted Godiva (Coventry Climax) 500gpm pump and a 40 gallon tank, plus an optional 17 gallon tank instead of the front passenger seat! Although never likely to seriously challenge Land Rover, the Gipsy had its fans especially Cornwall who purchased a large fleet but they were already Austin admirers thanks to their affinity for the Loadstar chassis.

Top Right: *A 1958 Land Rover 86 (NRT 283) supplied to Suffolk & Ipswich Fire Service and stationed at Bury St Edmunds as a general purpose vehicle.* Simon Adamson

Centre Right: *A rear view of one of Cornwall's Austin Gipsy light 4x4 pumping appliances. They had more than 30 of these, mostly long wheelbase, with one at each fire station. This 1963 example (153 TCV) served at Mullion.* Gary Chapman

Bottom Right: *The Austin Gipsy was popular for industrial use and this 1960 short wheel base example (VJB 748) was supplied new to Reeds' Colthrop Paper Mill at Thatcham in Berkshire. Now preserved in the New Forest, it ran alongside a 1956 Cumberland Bedford S/HCB WrT (RRM 564), which once served at Workington.* Simon Adamson

DENNIS

The success of the Dennis F12 and its smaller sister the F8 is part of Fire Service legend but the Guildford-based manufacturers were not content to rest on their laurels. Indeed, far from it, for in the post-war period they strove for further successes and in 1955 they unveiled their first fire appliance with a diesel engine, the F101. As has been noted earlier, the F7 was considered too long for the average user and less than 100 were produced. However by the end of 1951 its successor, the F12, had been ordered by no fewer than 90 British brigades, plus two from Ireland and four overseas. The biggest orders came from Middlesex (21), West Riding of Yorkshire (17) and Somerset (9). A total of 336 were built from 1950 until 1959 but some brigades wanted something smaller or something with a bigger water tank and thus the F8 and F15 models were born. In 1954 Dennis said that as the result of 'the closest co-operation possible between Brigade and Manufacturer' the Dennis F8 was available.

Two years earlier the first F8s had been produced for the Northern Ireland Fire Authority and known as the 'Ulster', but very soon the F8s were being snapped up in large numbers by rural brigades. Their 6ft 6in width was 12 inches less than the F12 and thus it was ideal for the country lanes, but it could still carry at least 300 gallons of water and a full complement of crew. Fitted with a Rolls Royce B60 petrol engine, nearly 250 were built until production finished in 1960. The larger tank came on the F15, which carried the 400 gallons needed to bring it up to full water tender specification. Although it was not built in vast numbers, it was nevertheless a popular choice for several UK brigades especially in Wales.

By the mid-1950s some brigades were anxious to run Dennis appliances with diesel engines rather than the Rolls Royce petrol engines fitted as standard. It was in March 1955 that Dennis announced that 'extensive trials of the new F101 12-litre diesel chassis are now complete.' The following month the appliance was being hailed as the new 'high performance' dual purpose appliance with a Rolls Royce diesel engine, Dennis 1,000gpm main pump, new 4-speed inverse drive gearbox, double reduction axle and constant flow hydraulic servo brakes. The London Fire Brigade purchased a fleet of 38 F101s from 1955 until 1960. Despite the attractions of the F101, many brigades still wanted to run petrol-driven appliances and so the F24 was conceived in 1957. It was powered by a Rolls Royce B80 engine but as the advertisement proclaimed, it was the first appliance in the field with automatic transmission, making it 'An Outstanding Success'. Derivatives of the F series carried on for years to come with a host of alternatives available with the main variants being between petrol or diesel, manual or automatic, or the type of appliance required.

Top Left: *A Dennis F15 water tender, with an unpainted aluminium body. This appliance (TTG 981) was supplied in 1957 to the Glamorgan Fire Service in Wales and served at Neath before passing on to the West Glamorgan Fire Service in 1974.* John Hughes

Bottom Left: *The Lincoln City Fire Brigade operated a 1952 Dennis F12 PE (JFE 415). It is seen in the company of a 1942 Austin K2 ATV (GLT 42), which had been converted into an emergency/salvage tender, at the scene of the fatal high-speed derailment at Lincoln Station in June 1962.*

Top Right: *The entrance to Wokingham Fire Station was too low to take a normal height fire appliance and so a specially constructed roofless Dennis F8 pump (UBL 465) was provided in 1960 by the Berkshire & Reading Fire Brigade. When a new fire station was opened, the appliance was rebuilt to a standard form in which it is now preserved. It was one of the last F8s built.* Simon Adamson

Centre Right: *One of the batch of Dennis F101s supplied to the London Fire Brigade in 1956 as dual purpose appliances - they could carry either a 50ft escape ladder or a 35ft double extension ladder. This one (SLW 167) served at Manchester Square.* Simon Adamson

Bottom Right: *An F24 water tender (HEF 264) which joined the County Borough of West Hartlepool Fire Brigade in 1957.* Simon Adamson

While it is acknowledged that the three major suppliers of fire engine chassis during the 1950s were Bedford, Commer and Dennis, there were a number of other makes who took a small share of the market. Amongst these firms was Leyland who had been a major pre-war supplier, but lost their way after nationalisation. However, they did make a real effort to appease the fire-fighter with a revolutionary fire appliance called the Leyland Firemaster. In trying to regain the custom they attracted before the war with their FK chassis, they began production of the Comet fire tender in 1949, in conjunction with Merryweather & Sons. Powered by a 6-cylinder Leyland petrol engine the Comets were bodied by several different bodybuilders. However the local authority brigades only purchased a small number of Comets, although several were exported, including an open model for Bombay. Despite their good looks, the Comets never appealed to the firemen and so Leyland decided to concentrate on commercial vehicles.

This could have been the end of Leyland's involvement, but in the late 1950s (following an approach from the Manchester Fire Brigade) Leyland Motors designed a purpose-built fire engine chassis. Using the main components of their Worldmaster bus chassis, the Firemaster appeared after considerable consultation with the Manchester Brigade. The new vehicle incorporated features that Manchester had pioneered, including a front-mounted Sigmund Pulsometer 900gpm pump, high pressure hosereels and semi-automatic transmission. Its 6-cylinder 0.600 diesel engine was horizontally mounted amidships and it carried a John Morris 55ft wooden wheeled escape ladder.

Despite all the hype surrounding its launch, the Leyland Firemaster was not a great success with only ten examples being built. Manchester had three pump escapes and an emergency tender, Glasgow had two pump escapes, Essex had a pair of pump/emergency salvage tenders and Darlington and Wolverhampton each had Magirus turntable ladders on Firemaster chassis. The last appliance was delivered in 1963 and most gave good service for nearly 15 years. Bodywork was completed by Cocker or Haydon with the exception of the Manchester's prototype (bodied by Carmichael) and their ET which was built by Smiths.

Dodge Brothers (Britain) Limited also made a brief foray into the fire engine market in the mid-1950s with their attractive pumping appliances. Owned by the American Chrysler Group, Dodge started building trucks at Kew in 1930; their vehicles at this time were based on 3- to 6-ton truck chassis. About 50 normal-control fire engines were built using the 5-ton chassis and in the October 1954 edition of *Fire Protection Review*, ambulance builders Herbert Lomas of Wilmslow, offered a diesel powered Kew Dodge water tender with its patent ladder loading gantry for a 45ft ladder. Supplied to Worcester City & County, it was equipped with a rear mounted Dennis pump and carried a Hathaway portable pump. Hampshire purchased six Carmichael built Kew Dodge F1s with 6-cylinder petrol engines, Devon had four and Breconshire & Radnorshire operated a few. Yet the vast majority, at least 20, were built by HCB for Essex who also had a pair of foam tenders built on a Kew Dodge chassis by Sun Engineering.

Top Left: *The Durham County Fire Service operated a 1952 Leyland Comet WrT (MUP 550) bodied by HCB; it is seen in the company of 1951 Commer 21A/Whitson WrT (LUP 289). Leyland, Dodge and Ford shared the same cab design.* Ian Moore

Centre Left: *The Carmichael-bodied Leyland Firemaster PE (WXJ 286) delivered to Manchester in 1959 and operated from London Road fire station. This picture shows the appliance in action at a fire at Strangeways and forms part of a set of colour postcards available from the Fire Brigade Society.* Bob Bonner

Bottom Left: *The only normal control Kew Dodge built as a pump escape was this 1955 appliance (NOT 405), with a Carmichael body, operated by the Hampshire Fire Service at Eastleigh.*

Top Right: *Austin-Morris (BMC) never made much of an impact on the UK fire engine market in the 1950s. A few Austin Loadstars and a number of Gipsies went to Cornwall followed by four Drake-bodied FFGs and an FFK4 in the 1960s, Nottinghamshire meanwhile had a few very unusual Wilsden-built K4 WrTs. Other brigades had a few pumping appliances but there were a number of special appliances supplied during the 1950s and 1960s badged as Austin, Morris or BMC. Among them was this foam salvage tender (999 JRA), one of three delivered in 1958/9 to the Derbyshire Fire Service on an Austin 503 chassis and built by Reeves. It was photographed at the Nottingham Road fire station in Derby.* John Shakespeare

Centre Right: *Ford made little headway in the fire appliance market until their 'D' series chassis, but in 1951 Fire Armour promoted their Firefly based on the Thames 3/5-ton long wheelbase chassis, which had a tank capacity of 500 gallons and a Coventry Climax 500gpm pump but the British brigades showed little interest. A few local authority brigades and the Civil Defence used Thames ET6 SWB control units usually bodied by Weymann or Marshall to a Home Office specification. The Ministry of Defence purchased 12 Fordson Thames 500E pump/foam tenders built by Wadhams in 1956, whilst the London Salvage Corps had four Fordson Thames 4D salvage tenders believed to have been bodied by Wood and Lambert including this one (NLB 101) supplied in 1953.* Simon Adamson

Bottom Right: *This Bedford TJ, supplied to the Moorsville Fire Board in New Zealand, should have been a popular appliance with rural brigades in Britain. Dennis Wilson who worked at Bedford in the 1960s writes. 'Although we were supplying forward-control versions of the TJ to British fire brigades, the Chief Officers would have nothing to do with the normal-control version. We were selling these into a variety of emergency service vehicle applications, notably ambulances, and it made sense for the county councils and municipal corporations to standardise on a common chassis. The TK was selling well, but we could not interest anyone in the 3- or 4-ton TJ in normal-control.'* Vauxhall Motors

MILITARY AIRFIELD APPLIANCES

Following on from the Thornycroft Nubian 4x4 Mark 5 crash tenders, first produced in 1951, a re-styled Mark 5A version was produced by Pyrene for the RAF in 1955 under a Ministry of Supply contract. The popular belief is that the 80 or so vehicles were actually sub-contracted to University Motors for their bodywork. Pyrene announced in the June 1955 edition of *Fire* magazine that it had received a contract for the supply of the Mark 6 crash tender which was described as 'an advanced design of fire fighting vehicle using an Alvis six-wheel cross country chassis with four wheel steering.'

The first of these appeared in 1957 and was based on the Alvis Salamander 6x6 chassis, which belonged to the same family as the Army's Saracen and Saladin. The Rolls Royce B81 in-line petrol engine was used in conjunction with a Wilson pre-selector transmission. It carried 700 gallons of water and 100 gallons of foam compound and the production run lasted until 1963. About 70 were built for the RAF and a handful for the MTCA. After some problems with the Mark 6 on overseas locations, they were all allocated to UK locations and the majority lasted until the late 1970s. During their lives various modifications were carried out making the derivatives Mark 6A to 6D. At the same time there was a requirement for a dual-purpose fire appliance to be used on military airfields.

The Thornycroft TFB 4x4 was chosen and it was supplied by Foamite although the bodies were built by Tecalemit in Plymouth. During the two years of production of the DP1, 1957/8, 200 were built for the RAF and MTCA with many going to overseas bases. They carried 700 gallons of water and 35 gallons of foam. A larger version, the DP2, followed in 1959 based on the Thornycroft TFAB 6x6 chassis with a Rolls Royce B81 petrol engine instead of the B80. The bodywork and fire engineering was supplied by Alfred Miles. The capacities were 1,000 gallons of water and 50 of foam. 60 were supplied to the RAF and Royal Navy and three to the Army and they lasted until the mid-1980s.

The Royal Navy were supplied with their own crash tenders which carried the designation TrkFFA/fld (truck fire fighting airfield). These were based on the Thornycroft Nubian 4x4 and 6x6 B81 powered chassis and were built by Sun Engineering. They had a separate Ford engine to power the fire pump. The capacities for the 6x6 were, 450 water and 100 foam and they could deliver 2500 gpm of foam.

Top Left: *Alvis Salamander 6x6/Pyrene Mark 6 A/C crash tender (23 AG 58) on the run with the RAF Fire Service at RAF Honington in Suffolk.* Simon Adamson

Centre Left: *One of just three Thornycroft Nubian 6x6/Miles DP2 crash tenders supplied to the Army Fire Service for use at the Army Air Corps airfields. This one (32 DM 32) was based at the AAC Middle Wallop near Andover, Hampshire.* Simon Adamson collection

Bottom Left: *Seen at the Royal Naval Air Station Lossiemouth (in Morayshire, Scotland), is a Thornycroft Nubian 6x6/Sun Engineering TrkFFA/fld crash tender (59 RN 22).* Simon Adamson collection

THROUGH THE 'SIXTIES

Whilst it is recognised that the 1950s saw considerable progress in fire appliance manufacture, the new technology of the 1960s enabled fire fighters to carry out their tasks with more reliable vehicles, pumps and equipment.

For much of the previous decade, there had been tight financial restraints in the wake of World War II. Yet as the new decade broke, research carried out by the manufacturers meant they were able to incorporate new ideas into their wares, and these appealed to the fire brigades which, at last, had some money available to ease the burden of their hard worked fire fighters.

Many brigades decided to change from petrol to diesel powered fire tenders although, in some cases, the petrol engine was to remain at the forefront of certain brigade's fleets for another 20 years. Automatic transmissions were becoming increasingly popular as the performance of the fire appliance was monitored closely.

Attendance times at incidents were important but at the same time the safety of the fire crews was not forgotten as body manufacturers looked at ways of improving the construction of coachwork.

Above: *Although about to be eclipsed by the TK, the Bedford TJ was a popular base for a variety of fire appliance applications for nearly 15 years. This J5 water tender escape (9113 UR) was supplied in 1963 to the Hertfordshire Fire Brigade. As can be seen it carries a variety of ladders including both a 50ft wheeled escape and a 35ft metal double extension.* Vauxhall Motors

Multi-pressure pumps and high-pressure hosereels were now available whilst the 50ft wheeled escape ladders were beginning to disappear and were being replaced by 45ft (13.5m) metal triple extension ladders. Another innovation of the early 1960s was the hydraulic platform, which was hailed by manufacturers Simon Engineering as "A big new advance in firefighting" and there is no doubt that the Snorkel seriously threatened the traditional turntable ladder in the aerial fire appliance market.

Dennis continued to gain orders from far and wide whilst AEC, Commer, Dodge and Ford retained a share of the market but, as the new decade dawned, they were all about to be confronted by their most serious challenger yet in the form of the Bedford TK.

AEC (Associated Equipment Company)

To use a phrase appropriate for this publication, it would be true to say that AEC 'didn't set the world on fire' when it came to selling chassis for fire engines. Yet, on the other hand they did have a faithful following which enabled them to maintain a significant niche in the market. Their Regent III chassis had proved its worth through the early part of the 1950s, but by the middle of the decade there was a need for a replacement incorporating new features.

Merryweather, who had been closely involved with AEC in developing their fire tender range, offered the Marquis in 1955, which they described as 'the finest machine on the road today'. Since Maudslay became part of the AEC fold in 1948, fire engine chassis had been constructed at the Castle Maudslay plant at Alcester in Warwickshire. It was a Maudslay chassis that was chosen for the Merryweather Marquis.

An advertisement boasted that water tenders had been sold to Surrey and Dewsbury and two appliances were to go to Edinburgh in the South Eastern Fire Brigade. However their confidence was soon shattered as only six were sold due to various problems, including the fact that it was very difficult to drive and it certainly wasn't the best looking fire engine in the world! It was back to the drawing board, but very soon Merryweather offered fire tenders using the highly attractive AEC Mercury and this was a winner!

The first of nearly 200 Mercury appliances appeared in 1957 and the line continued well into the 1960s before being superseded by the equally successful Mercury TGM chassis with the Ergomatic cab. The low centre of gravity on the AEC Mercury chassis was ideally suited for a turntable ladder adaptation and just over half the total were used to carry Merryweather four-section hydraulically operated 100ft ladders constructed in light alloy. All had full crew cabs with the exception of a TL supplied to Stockport, which had a single cab.

Surprisingly three Magirus and two Metz TLs were also carried on AEC Mercury chassis. Fitted with an AEC AV470 7658cc diesel engine, there were a number of AEC pumping appliances constructed with the main orders coming from Glasgow (14), London (11) and the South Eastern Area of Scotland (10). The vehicles were better suited to urban rather than rural areas and this was reflected in the number of city and town brigades buying examples. Specials included three hydraulic platforms and 12 emergency tenders.

The AEC Ergomatic cab made its first appearance on a fire appliance in 1967 and, using the Mercury TGM chassis, totalled about 100 when the range came to an end in 1974. The four-section Merryweather TL accounted for 20 (all except two for Dorset had full crew cabs), there were four HPs and about 10 specials but the rest were pumping appliances with Dublin (9), Sheffield (7) and Hertfordshire (7) being the main buyers.

In turn the Associated Commercial Vehicles group, in which AEC was the main player, become a part of the British Leyland empire in 1962. After this it gradually lost its individuality with the Ergomatic cab finding its way onto several Leyland products, before the AEC marque disappeared in the mid-1970s. Leyland, who joined forces with BMC in 1968 to form British Leyland, produced about a dozen Beaver pumping appliances for Plymouth and St Helens, whilst the county boroughs of Derby and Hartlepool each brought a Lynx pump. With the exception of a few specials, Leyland departed from the fire engine scene until the mid-1980s.

Top Left: *Typical of the AEC Mercury/Merryweather TLs is this 1961 example (194 SBB) which was new to the Newcastle & Gateshead Joint Fire Service and served at Pilgrim Street (Newcastle) and then at Gateshead under the Tyne & Wear Metropolitan Fire Brigade.* Simon Adamson

Top Right: *Carmichael built the body on this 1962 AEC Mercury (3 BBL) which served the Berkshire & Reading Fire Brigade. As it carried a 45ft triple extension ladder, it was classified as a water tender ladder (WrL).* Simon Adamson

Centre Right: *Another AEC Mercury/Carmichael appliance for Berkshire & Reading was this unusual special unit (XJB 999) which was supplied in 1961 and was stationed at Dee Road (Reading). It combined the duties of foam tender, salvage tender and control unit.* Simon Adamson

Bottom Right: *The Ergomatic cab, with all its added refinement and consideration toward driver comfort (hence the name ergomatic) was used on numerous AEC fire appliances. It set a standard for others to follow, including Leyland who became associated with AEC as part of British Leyland. Here an ergomatic cab is seen fitted on a Leyland Beaver chassis (LJY 999H), which was built in 1970 for the City of Plymouth Fire Brigade. It served at Greenbank latterly under the Devon Fire Brigade.* Alan Batchelor

BEDFORD PROGRESS

As the new decade dawned, Bedford were offering four chassis for fire engine application. Although the long established S type and the closely related C type were on the way out, the J type was destined to carry on for many years as several fire brigades chose the forward-control version as the basis for water tenders and some specials. The 4x4 R series was also available for several years. But within a few months the prototype Bedford TK was to appear and that opened the floodgates with chief fire officers all over the UK and abroad wanting to examine this revolutionary vehicle. The 6-ton TKEL and 7½-ton TKG were destined to become the most popular British fire engine chassis ever built, with literally hundreds supplied over the next 20 years or so. The early Bedford TK fire engines were fitted with Bedford 300 petrol engines and it was not until the mid-1970s that the 500 diesel engine became the standard power unit.

Given the undoubted success of the TK, it was strange that a number of brigades still insisted on the TJ! Dennis Wilson, who worked for Bedford, writes, 'The position was bizarre because we could not see the reason why the fire brigades wanted us to convert the TJ to forward-control, when the TK was designed for this role. Our concept was to supply the TK as forward-control and the TJ in normal-control versions. The TK was more than admirable for lighter duties, but fire brigades only seemed to consider the TK for heavier applications. It sold very well, so we didn't complain, but it has to be said that the TJ forward-control was very much a compromise and the fire service did not get the best deal for their money.' He adds 'the sales of TK were so easy to achieve, that we never had real problems here, but I remember taking one to demonstrate in Holland where we'd had some limited success with 4x4 chassis, but I was completely surprised when their Chief Vehicle Inspector took one look at the TK, and said 'it's just too ugly'.

Although several bodybuilders, including Carmichael, Dennis (who had taken over Alfred Miles), Smiths and Pyrene, at some time built bodies on the Bedford TK, it was HCB Engineering, and then HCB-Angus, who had the biggest involvement with the Marque.

With the approval of Bedford, HCB had worked on the forward-control versions of the D and J types and on uprating the TK petrol engine (though the development work was actually done by Janspeed of Salisbury on behalf of HCB-Angus). Here we must take a slight diversion to look at the formation of HCB-Angus which came about in 1964 when HCB Engineering became associated with the rubber company George Angus & Co Ltd, who in turn owned the London-based Fire Armour. At once all building of fire appliances by Fire Armour was transferred to the HCB-Angus plant at Totton and the FA brand name of Firefly was passed on to appliances built by HCB-Angus and lasted until the 1970s. At this time they were part of the Angus Fire Armour Division of the Dunlop-Angus Industrial Group.

The first special fire appliance to use a Bedford TKEL chassis appeared in 1961 and within a few years all types of specials had been built. These included hydraulic platforms, emergency tenders, salvage tenders, foam tenders, control units, water carriers and canteen vans, all utilised the chassis; meanwhile the heavier TKGL chassis was used to mount turntable ladders.

Top Left: *Dating from 1962, HCB built this Bedford RLH pump foam tender (99 PHY) one of a pair supplied to the Bristol Aircraft Limited Fire Brigade for use at Filton Aerodrome. Vauxhall Motors*

Top Right: *A publicity photograph of a Merryweather/Bedford TK foam tender, which apparently was built for 'municipal customers in January 1964.' It looks identical to several appliances built for export although some UK refineries operated similar appliances. Vauxhall Motors*

Centre Right: *The Breconshire & Radnorshire Joint Fire Brigade purchased this Bedford TKHE (FFO 606K) in 1971; powered by a Perkins V8 diesel engine, it was fitted with Simon SS70 booms and had bodywork by HCB-Angus. From new it was stationed at Llandrindod Wells (under Powys from 1974) until disposal in 1996 and, after refurbishment by Angloco, it joined the BAA Fire Service fleet at Heathrow Airport. In August 1983 it was badly damaged (by radiated heat from a boil-over) during a severe oil tank fire at the Amoco Oil Refinery at Milford Haven, Dyfed. It therefore had to be returned to Simon Engineering at Dudley for major repairs. Barrie Lowe collection*

Bottom Right: *One of the last appliances to be built by HCB Engineering before they became HCB-Angus in 1964, was this Bedford TKEL combined emergency and salvage tender (JEK 999). It was supplied to the Wigan County Borough Fire Brigade, and on re-organisation passed to the Greater Manchester Fire Service before disposal in 1975. Vauxhall Motors*

Top Left: *This publicity photograph was taken at the Chief Fire Officers' Conference in 1964 and shows a TKLG chassis with Merryweather body and 100ft TL. The 1962 Manual of Firemanship said it: 'consists of a main ladder and three extensions, giving a total extension of 100ft. The strings and rounds are constructed from rolled steel sections of special form and the trussing from tubular steel, welded to the strings. A unique feature of this ladder is that the fulcrum frame has a platform, on which the operator stands, so that he is carried round with the ladder when it rotates and thus is always in a convenient position for operating the controls. The turntable is mounted on rollers. A PTO from the road engine is employed to transmit the motive power to the ladder drive.' Five TLPs (fitted with pump) like this were delivered to UK brigades in 1963/4.* Vauxhall Motors

Below: *At the scene of a road accident on the A35 near the Cat &Fiddle at Hinton, in February 1969, is New Milton's Bedford TKEL WrT (895 DHO). This was the first Bedford TK to be built by the Hampshire Fire Service Workshops in 1963. It was fitted with a Dennis No 2 pump salvaged from a wartime trailer unit. In 1979 it was refurbished and donated to the Fire Service National Museum Trust for their proposed National Museum of Firefighting, which is likely to be established in Northamptonshire.*

COMMER KARRIER DODGE

The Rootes Group appliances looked quite dated as the 1960s arrived, but there were still devotees to both makes. For example Berkshire & Reading bought 12 Karriers, whilst Buckinghamshire went for Commers with bodywork built by T & G Button. This Welsh coachbuilder from Pontllanfraith near Blackwood in Monmouthshire was also accredited with building fire tenders for their local brigade. A bodybuilding firm that had long been associated with the Rootes Group chassis (Alfred Miles of Cheltenham), was acquired by Dennis Brothers in 1962 and although Miles bodies continued to be produced for a short time they were badged as Dennis products. A few former Miles employees (perhaps viewing the opportunity to carry on working with Rootes) purchased a business named Healeys. However, records show that fire tender orders were virtually non existent and it seems as though only a pair of Commers were bodied for the Northumberland brigade.

In 1963 the Commer VAK was introduced and given a completely new cab, the 6- and 7-ton chassis from this range were used for fire appliances. One of these formed the basis of the first hydraulic platform supplied to a UK brigade (Monmouthshire), on which Simon Engineering provided 65ft booms and Button did the bodywork for a revolutionary concept which would find its way into the majority of UK fire brigades. Meanwhile early examples of the Commer VAK water tender (fitted with Rootes 6P.290 vertical petrol engines and were equipped with Coventry Climax 500gpm pumps) went to Perth & Kinross and Caernarvonshire. A big order for Kent followed which they designated the K2.

After its limited success in the mid-1950s, Dodge followed the normal-control 134 with a forward-control version. One of the earlier models had been equipped with a Hayward Tyler pump, but fitting Godiva/Coventry Climax or salvaged Dennis No 2 pumps was the normal practice. Again it was Devon and Essex who provided the bulk of the orders for the forward-control Dodge, with HCB constructing the bodies using their own standard design (sometimes totally unpainted). Production ceased in 1960 and the Dodge fire engine disappeared for several years. At the end of the decade Commer and Karrier joined Dodge under the Chrysler UK umbrella.

Top Right: *Victor Healey of Cheltenham built this Commer 86A WrE (ATY 401B) for Northumberland who used it at Ashington and then Wallsend.* Simon Adamson

Centre Right: *This 1969 Karrier Gamecock/Carmichael WrT (TBL 485G) was supplied to Berkshire & Reading, it is pictured here attending an incident whilst stationed at Crowthorne.* Simon Adamson

Bottom Right: *This Dodge 123A (YOD 982) had HCB bodywork and PTO, a Godiva pump and two-speed axle and was used as Bovey Tracey's WrT from 1958. At this time Devon County Fire Service favoured the unpainted body and there is no red paint in sight on this appliance!* Roy Yeoman

LIGHTWEIGHT APPLIANCES

The new phenomenon of pedestrian precincts gave some rise for concern, because the limited access could delay full sized fire engines from getting to the scene of a fire. However, as small fire tenders were already used in factories with restricted access and in towns where narrow streets caused problems, the lightweight appliance concept was thought to be the answer. In 1966 the Newcastle & Gateshead Joint Fire Service purchased two mini Leyland fire tenders which were designed to operate in tandem. Between them they carried the same equipment and crew as one full size appliance although they relied on hydrants for their water supply. The chassis was a derivative of the Standard Atlas and the bodies were built locally by the local main dealers, Minories. They were equipped with transverse Coventry Climax pumps, one carried a metal 45ft ladder and the other a wooden 35ft ladder.

As well as their use in locations with difficult access, lightweight appliances could also be purchased more cheaply than a full sized machine and the driver did not need a heavy goods driving licence. Small van type vehicles such as the Ford Transit and Bedford CF were fitted with 40 or 80 gallon water tanks and hosereels and were used as 'first strike vehicles', but they were always backed up by a predetermined attendance of full size appliances. There were also concerns about fire spread in multi-storey car parks but in the event, the inclusion of adequate fire precautions such as dry risers, lessened the risks and the need for small fire appliances was no longer thought to be a necessity. However the cost savings remained and brigades continue to use this type of appliance to this day.

Another need for a small appliance was as a rescue tender where a crew of two could make good progress through heavy traffic to a road accident or similar emergency. The rescue tender was in essence a small version of the emergency tender although some brigades use the ET designation for both types of vehicle. The *Manual of Firemanship* describes the ET thus: 'These appliances are specifically designed to carry a wide range of special-purpose equipment, both for firefighting and for the many types of special service that fire brigades are called upon to provide.'

Top & Centre Left: *These two views clearly show the different ladders carried by the pair of Leyland Atlas mini pumps with Minories bodywork, bought by Newcastle & Gateshead for use in pedestrian precincts. Above is the newer one (LVK 439E) with a 35ft wooden Ajax ladder whilst below we see a picture of the older appliance (KYK 694D) with a 45ft metal Lacon ladder.* Simon Adamson

Bottom Left: *Although the Berkshire & Reading Fire Brigade called this Commer Walk-Thru K30 (ORX 206F) an emergency tender it was actually a rescue tender. Supplied in 1967 it was operated from Caversham Road, Reading. Tools and various types of lifting gear were also on the inventory. In vehicles such as this, equipment was stowed in racks inside the vehicle with access through the rear doors or the cab's wide sliding doors.* Simon Adamson

The light Rescue Tender could be manned by two firemen, and in the 1960s its carried a whole array of equipment including oxy-acetylene cutting equipment, but oxy-propane cutting sets and hydraulic rescue sets using pumps and rams were soon in regular use. There was also a need for small general purpose vehicles to carry personnel and equipment and standard production types were generally used. One interesting vehicle was Devon's Austin Mini Moke.

As stated, 4-wheel drive capability was also important to brigades with large areas of heaths or moorland and also to the airport fire service which used this type of appliance as a 'rapid intervention vehicle' (RIV). Land Rover was the main choice for many users, with the Series III 109in being used by local authority, industrial and military brigades. Many of the industrial brigades preferred the Land Rover appliance fitted with a 500gpm fire pump, which was effectively a scaled-down version of a full size water tender. Local authorities had some of this type but, they mainly used their vehicles as hosereel tenders. The forward-control 110in Land Rover was also adapted as a rescue tender or pump and it is estimated that in excess of 360 Land Rovers were operated by local authority brigades in the period concerned.

Land Rovers were also very popular with the military brigades, but. the RAF initially used converted Willys Jeeps. Then followed the Land Rover 86in and 88in models, before the 109in was adopted in 1962. Foamite (taken over by Merryweather in 1965) produced an airfield crash rescue tender (ARCT) which, according to RAF records had three main functions; immediate rescue from crashed aircraft, extinguishing aircraft wheel-brake fires and escorting aircraft to dispersal points as a precaution against taxing fire hazards. About 200 were built and they carried dry powder and portable extinguishers plus rescue tools. They were followed by the TACR1 (truck airfield crash rescue) 1-ton Land Rover 109 of which 80 were built by HCB-Angus for the RAF, Navy and Army.

Domestic risks were covered by Land Rover forward-control light pumps with 15 Carmichael Redwing FT6's followed by 70 from HCB-Angus. They carried 115 gallons of water and a crew of four. In 1970 the first Range Rover fire tender (The Carmichael Commando) was pioneered by Carmichael; it used a 6x4 configuration with the trailing axle a Carmichael patent. It carried 200 gallons of pre-mixed foam and 100 kgs of BCF and become the standard RIV for all types of airports.

Top Right: *Despite being owned by the RAF Fire Service, this Land Rover 109/Foamite ACRT carried the number COU 774C as it was used by civilian staff at The College of Air Training in Hamble.*

Centre Right: *Merryweather called their Land Rover FC110 conversions 'Fire Wardens'; this one (BFS 330B) was delivered to the South Eastern Fire Brigade (Scotland) and operated from Bathgate as an RT but is seen in the Lothian & Borders fleet.*

Bottom Right: *The prototype Range Rover 6x4/Carmichael Commando RIV (YVB 152H) which was used as a demonstrator; it is seen here at the 1970 Farnborough Air Show but it is now operational at the Marshalls' Cambridge Airport. All three pictures, Simon Adamson*

FORD OF BRITAIN

Despite their post-war dominance of the car market, the Ford Motor Company made very little impact on the British fire engine market. Despite some incursions into this specialist field, Ford never really undertook any serious marketing of their products until the arrival of the Transit and the D Series in the mid-1960s. Until then just a handful of Thames Traders and a spattering of 15cwt Thames vans had been used, mainly as special appliances. The dozen or so Traders had a 138in wheelbase and were powered by a 114 bhp 6-cylinder petrol engine; they included four foam carriers for the Glamorgan County Fire Brigade in Wales and salvage tenders bodied by Wood and Lambert for the London Salvage Corps.

There were four pumping appliances, one of these was used by the Ford Motor Company's own brigade at Brentwood, and carried the standard Alfred Miles body. Two saw service with Irish brigades and of these Wexford had a Fire Armour bodied example, whilst Kerry had one bodied in Ireland by Roberts and it still survives. The City of Oxford Fire Brigade operated one of the last Thames Traders to be used as a fire appliance with a 1963 example built by HCB as an emergency tender.

About this time the D series was launched and the first ones appeared as fire appliances in 1967. According to Ford's records, the firm now made a significant attempt to market their new commercial vehicle chassis and their Transit chassis-cowls for 'new applications'. A significant audience was seen in the local authority, municipal and government departments. As Ford had already supplied the Government with chassis cabs on the Ford Thames 400E for Civil Defence ambulances and control vehicles, they knew that here was a highly profitable sector of the 'special market'. The story of which is told in the **Nostalgia Road** book *NHS Ambulances - The First 25 Years*.

Over the next seven years more than 250 were adapted for fire appliance use in the UK, of which only a small proportion of these were specials. Ford made a huge impact on the pumping appliance market with a variety of bodybuilders receiving orders; these included Merryweather who built 20 Marksman whilst Cumberland ordered a Saro and Monmouthshire a Button. Northumberland used three local firms Hollowell, Killingworth and North Eastern to build bodies on various D series chassis.

The City of Gloucester Fire Brigade purchased a D600 pump escape built by Pyrene, who were normally associated with building refinery foam tenders. The County of Gloucestershire were attracted by the basic 'lo-cost' HCB-Angus body, which had a normal crew cab with rear cab and bodywork as a separate unit! The D600 had a 134 in wheelbase and was supplied with either a 6-litre turbo-charged diesel engine (140 bhp at 2400 rpm) or 300 cubic inch petrol engine (130 bhp at 3600 rpm). The larger D1617 chassis was used for a number of hydraulic platforms, some being delivered into the 1980s, whilst the most unusual D series fire appliance was a 4x4 D800, which was bodied by Locomotors as a hose layer for the London Fire Brigade in 1971. The Ford D series chassis (from the D300 to the D2817) were adapted for fire engine use and continued for 15 years.

The Transit was found in many forms throughout the fire service, although the main buyers were the industrial brigades. The models in the 1960s were based on the 30cwt van and usually were powered by a Ford V4 petrol engine. The Transit's 118in wheelbase enabled it to be manoeuvred easily in tight locations, yet it was still able to carry at least 100 gallons of water and a 400gpm Coventry Climax pump plus a crew of four. The pump was mounted at the rear and the first-aid hosereel was positioned above the tank and accessed through the lift-up tailgate. HCB-Angus demonstrated their example in 1967 and other builders soon followed with similar versions of a Transit light pumping unit. It appeared in various other guises including rescue tender, salvage tender, control unit, canteen van and breathing apparatus tender.

Top Left: *A pre delivery photograph of the Ford D600 pump escape (FFH 960K) built for the City of Gloucester Fire Brigade by Pyrene. This was one of only a handful of local authority pumping appliances built by Pyrene who were well known for producing foam tenders. The appliance carried a 50ft Merryweather all steel escape ladder which had a weight of 10½ cwt. This particular appliance was the first one, which was preserved by the author. Its strange appearance earned it the name 'Trumpton' when it was in service.*

Top Right: *The City of Oxford used this 1963 HCB-bodied Ford Thames Trader (288 RFC) as an emergency tender and allocated it to the Rewley Road fire station.* Simon Adamson

Centre Right: *The London Fire Brigade operated a fleet of 11 Ford Transits as breathing apparatus control vans supplied between 1971 and 1973. A similar vehicle was used as a high expansion foam unit (Hi-Ex FoamU). This BACV (JLT 41K), which was fitted out by the LFB workshops, was based at Stratford and lasted until 1980.* Fire Training

Bottom Right: *An example of the HCB-Angus 'Lo-cost' body on a Ford D600 WrT supplied to Gloucestershire. Its spartan appearance made it unpopular with crews and after 16 were delivered, the brigade reverted to a more conventional style of body. This one (WDF 803J) served at Stroud.* John Hughes

READY FOR THE 'SEVENTIES

As the 1970s approached, the range of fire tenders and equipment available was increasing as advances in technology were being used to assist the firefighter in all spheres of his hazardous task. At the same time fire appliance manufacturers strove to get better performances from their vehicles but without increasing their costs or lessening the prime importance of crew safety.

The name of Dennis had become synonymous with fire engines worldwide, and the firm claimed to have their appliances operating in 46 countries. Always looking for ways to keep their prices at a competitive level, they introduced their D series specifically designed for rural use. It was only 7ft wide and was powered by a Jaguar 4.2 petrol engine. The F series had Rolls Royce petrol and Perkins diesel engine options, but the aluminium-covered ash frame made it expensive and the series came to an end in the mid-1970s. In 1963 they introduced the Delta chassis, with the driving position set ahead of the front axle thus allowing the cab to be built lower which could accommodate the booms of an HP; although only a limited number were produced, the later Delta II found many buyers. Despite a full order book, Dennis was struggling to survive and, in a take-over not welcomed by the board, it became a part of the Hestair group in 1972 for a meagre £3.4 million. After selling off some loss-making sections and its Guildford factory (it rented back the proportion it needed), Dennis-Hestair concentrated on selling fire appliances and trucks to local authorities.

The old firm of Merryweather & Sons, also faced a serious challenge, this time from the two West German turntable makers, Magirus and Metz. Their products were basically similar to the ladders supplied by Merryweather, but many brigades chose to buy from the Continent rather than buy British. Cost implications and standard of product no doubt were the prime factors but whatever the reason, the B type ladder was on its way out and it spelt the beginning of the end for this famous fire apparatus manufacturer. It would be prudent to mention here the business of G & T Fire Control who were based at Northfleet in Kent; this company reconditioned Merryweather turntable ladders, which were then re-used on new chassis. It should also be noted that Magirus introduced its DL30HF turntable ladder in the early 1970s, and this ladder, complete with its rescue cage, was to prove very popular throughout the world.

A make not covered so far is Albion Motors of Scotstoun, Glasgow, who provided chassis for fire tenders before World War II especially for mounting Merryweather TLs. Leyland bought Albion in 1951 and in the 1960s the Reiver, Claymore and Chieftain chassis were used for seven special appliance applications. Chieftain pumping appliances were built in conjunction with Carmichael using their Vista-Vue cab and were christened the 'Fire Chief'. They were powered by a Leyland 6.5 litre L400S diesel engine (125hp) and were equipped with a Gwynne 500gpm pump; 32 were built by Carmichael, including seven for Leicestershire and six for Manchester who also had six with Cocker bodies and four from HCB-Angus; the grand total was 50 pumps and specials.

Top Left: *Eastleigh fire station in Hampshire with Dennis D WrT (AOT 239J), a 1967 Bedford TKEL foam tender (ECG 767D) and a 1954 Commer 45A (KDG 112). This appliance was originally supplied to the Gloucestershire Fire Service and carried a Miles built WrT body but was written off in an accident; it was purchased in 1962 by Hampshire for spares but it was subsequently rebuilt in their workshops as a PE.*

Bottom Left: *A Magirus 100ft TL mounted on a 1964 Commer C7 chassis with bodywork supplied by David Haydon. This TL (BHH 385B) was supplied to Carlisle and passed to Cumbria in 1974; the ladders were later put onto a Dodge G16 in 1980. Simon Adamson*

Top Right: *A Metz 100ft TL mounted on a 1958 Dennis F27 chassis, (250 AUP) seen at Stockton. It was new to Durham County but was owned by three subsequent brigades, Teesside, Cleveland and Humberside. Simon Adamson*

Centre Right: *The Northern Ireland Fire Authority purchased Dennis fire appliances throughout the 1950s and 1960s before turning to Bedford and Dodge in the 1970s. This 1964 Dennis F28 (9430 KZ) was originally stationed at Coleraine as a PE. Barrie Lowe*

Bottom Right: *New to the Angus Fire Brigade, this Albion Chieftain WrL (MTS 722J) was bodied by Carmichael with their Vista-Vue cab. It served at Forfar and is seen after transfer into the Tayside Fire Brigade fleet. Ken Reid*

Until the mid 1960s ERF had shown no interest in the fire engine market, but it was the arrival of the Simon Snorkel hydraulic platform that propelled them into the limelight as one of the main suppliers of chassis suitable for carrying HPs of all proportions. The first of these was to the City of Lincoln delivered in 1967, but ERF had already supplied their first fire appliance chassis the previous year. This was a pump escape for the Newcastle & Gateshead Joint Fire Service, fitted with a body built by HCB-Angus and powered by a Perkins V8 510 diesel engine. In the period up to 1974, ERF went on to provide chassis for about 130 fire engines. Of these there were 74 HPs, 52 pumping appliances, 2 ETs and 2 TLs. HCB-Angus bodied 94 of them and Jennings was responsible for 26; Fulton and Wylie did the bodywork for eight HPs for Scottish brigades, whilst the Stockport coachbuilders Kirbell built two water tenders for their local brigade.

The 85ft and 70ft HPs were mounted on a 16-ton chassis with a 17ft wheelbase, which were designated 84RS or 84PS depending on whether a Rolls Royce petrol or Perkins diesel engine was supplied. The 50ft HPs came on a 13.25-ton chassis with 12ft 6in wheelbase, which could also be used for pumping appliances as could the 12.75 ton 11ft 10in version. The smaller chassis were designated 84RF or 84PF. About this time ERF bought out the long established coachbuilders J. H. Jennings & Co. who, like them, were based in Sandbach, Cheshire; they had been responsible for the double ended cabs which became a feature of many of the ERF fire appliances.

With the increasing numbers of calls and varying kinds of incidents, the fire brigades were increasing their types of special appliances. As no two brigades were alike in their requirements, it was left to each chief fire officer to specify the types of appliances that were needed to fulfil his brigades obligations. The Home Office laid down the basic guidelines but even neighbouring brigade fleets could vary considerably. Obviously pumping appliances were essential but sometimes the smaller brigades relied on their neighbours to provide them with aerial appliance cover. Specials such as water carriers, hose layers, breathing apparatus control units and foam units were only found in some brigades. Breakdown lorries also had a limited appeal and certain brigades used them as heavy emergency tenders whilst others had no use for them at all. The subject of breakdown vehicles in fire brigade use will be covered in more detail in a forthcoming **Nostalgia Road** publication about towing vehicles.

As was observed earlier, the bulk of the military airfield crash tenders were built on the Thornycroft Nubian chassis and it was the same story with the civilian airport fire engine fleets. The tried and tested Nubian could not be bettered and specialist crash tender builders like Pyrene, Fire Armour and Sun Engineering had a strong hold on the market in the early 1960s. Carmichael built their first Nubian Major TMA-300 6x6 crash tender in 1966; using a 20-ton chassis, the Jetranger was powered by a Cummins V8-300 diesel engine with a 5-speed semi-automatic gearbox. It carried 1500 gallons of water and 150 gallons of foam and its roof-mounted monitor could deliver 7000 gpm of foam to a jet length of 200 feet. About this time Pyrene announced their Protector range, whilst both Merryweather and HCB-Angus tried crash truck building but with limited success in the UK, although the latter did sell a number of crash tenders to the Royal Navy.

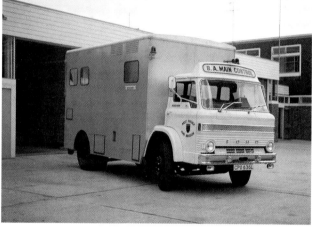

The Thornycroft Nubian Major 6x6 was the preferred chassis for a fleet of crash tenders supplied to the British Airports Authority Fire Service. The initial batch of eight were bodied by Locomotors, whilst Gloster Saro built 15 in two batches. Compared with Thornycrofts, Bedfords had been used infrequently as crash tenders, although about 20 RLHZ types were supplied to municipal airports and a similar number were used as airfield rescue tenders. This chassis was chosen as the basis of the RAF Mark 8 crash tender, and about 50 were built by Pyrene who had earlier completed the same number of Thornycroft Nubian 6x6 Mark 7 crash tenders .

Top Right: *This 2,000-gallon water carrier (YAF 725J) was built in 1971 on a Leyland Laird chassis for Cornwall and stationed at Bodmin. When it came out of service in 1987, it was refurbished by Action Water and donated to Sierra Leone, West Africa where it is still used to ferry drinking water to drought stricken areas. Gary Chapman collection*

Centre Right: *The West Sussex Fire Brigade had three Ford D2817 breathing apparatus support vehicles built by Contact. This one (CPO 636L) served at Horsham from 1972 until 1996. Martin West*

Bottom Right: *British Airports Authority Fire Service Thornycroft Nubian Major 6x6/Gloster Saro crash tender (MHM 961L). Before going into service at Heathrow Airport it was displayed at the 1972 Farnborough Air Show. Simon Adamson*

Above: *Berkshire & Reading operated this Orbitor 72ft HP on a Dodge K1050 chassis with Carmichael bodywork (JMO 115K) from the Caversham Road fire station in Reading.* Simon Adamson

After a break of several years, Dodge, now a part of Chrysler UK, made a fresh attempt to gain a substantial share of the fire engine market and this time made a considerable impact with its K Series. Although launched in the mid-1960s, it was the end of the decade before the K850 was available for fire engine applications. The Perth & Kinross Fire Brigade took the prototype with HCB-Angus bodywork and soon brigades throughout the UK were adding the Marque to their fleets. Powered by a Perkins V8 510 diesel engine (185 bhp) coupled to a 5-speed synchromesh gearbox, the K850 was equipped with a Godiva UMP 500gpm pump. Its measurements were 22.5 ft long, 7.5 ft wide and 9.5ft high. Unipower of Perivale, Middlesex, brought the chassis up to fire engine specification and more than 100 were sold in the first three years of production with K1113 being adapted for most pumping appliances. The K1050 chassis was used as the basis for nine Simon Snorkel HPs but the Berkshire & Reading Fire Brigade chose a different make of HP, the 72ft Orbitor from Finland for which Carmichael held the UK concessions. This platform never challenged the Simon Engineering range and only a handful were sold in Britain including one for Bradford mounted on a Bedford TKM chassis.

Throughout this book progress in the manufacture of fire appliances has been discussed but advances were made in many other spheres of the firefighter's equipment. Hose for instance was originally cotton-based canvas and it had to be scrubbed and dried after use or else it deteriorated. Station towers were then used as much for hanging hoses up to dry, as for drilling purposes or carrying call out sirens (which were also being replaced by the 'bleeper' paging alerters). Hose made from man-made fibres gradually replaced canvas and could be restowed on the appliance wet. Couplings, originally brass or gunmetal began to be made of light alloy with considerable weight saving. Branches and nozzles were likewise made of alloy instead of brass and became much more sophisticated with on/off controls and the ability to change from jet to spray. At the end of World War II, standard ladders were the two-section extension, made of timber (typically 30ft or 35ft), and the 50ft wheeled three-section escape usually made of timber.

Occasionally steel was used, but weight-saving led to the use of light alloy ladders and the replacement of escapes by three section 45ft ladders (Merryweather and Lacon being the pioneers). The demise of the wheeled escape was hastened by the difficulty in obtaining the 5ft diameter carriage wheels on which the ladders were moved about. Self-contained breathing apparatus became common between the wars, and oxygen gave longer duration than compressed air but was less convenient to wear and service. It was generally used as a last resort, but in the 1950s and 1960s more emphasis was placed on the need for BA wearing, but even into the 1970s some pumps carried no BA at all.

Before the war, the hand operated bell and a loud voice were the usual audible warning systems and headlamps or occasionally flashing lights (red, amber or green) were used. Post-war, the electric bell and siren were common, and continental type two-tone horns became standard from the 1960s following government research; actually it was found that two-tone horns were better than bells in terms of effectiveness but were outclassed by four-tone horns and sirens. Following problems experienced at the scene of motorway crashes, especially on the then recently opened M1, revolving blue lights replaced the repetitive amber flashers to give motorists more warning of an impending obstruction; but although the lights were effective at night they proved to be unsatisfactory in daylight. Often rescue tenders were painted white so they could be seen more clearly and at the start of the 1970s emergency services were experimenting with high visibility 'day-glo' paint and striping to highlight their vehicles.

Fire engines had to carry an increasing amount of equipment as the role of the brigades changed to provide a general rescue service. Items such as hydraulic jacking and spreading equipment and general rescue tools had to be carried firstly on emergency tenders and breakdown lorries and later on smaller rescue tenders and standard pumping appliances.

In the late 1940s low-pressure pumps, generally 350-500gpm were standard. The Americans had pioneered high pressure pumps in World War II, which gave a water fog (very fine spray) developed for airfield firefighting but it was found to be a very effective way of fighting internal building fires. Early examples of this system, applied to UK fire engines, featured separate high-pressure pumps feeding high-pressure hosereels with fog guns. The main pump continued to supply low-pressure jets through the conventional hoses. But soon the manufacturers produced multi-pressure pumps, which could be switched from low to high pressure. However many brigades stubbornly ignored multi-pressure pumps and it was not until the 1980s that they became universal.

Top Right: *Fife Fire Brigade's all red-livery, Dodge K850/HCB-Angus WrL (VSP 983L) served at Glenrothes.* Ken Read

Centre Right: *The Holland County Fire Brigade in Lincolnshire preferred an all white livery for their Land Rover 109/Carmichael RT (VDO 99K) which served at Sleaford.* Martin West Collection

Bottom Right: *One of the Glasgow Fire Brigade's four Magirus Deutz 150D WrEs with bodywork by SMT. This one (NGE 47F) served at Central.* Alistair MacDonald

Appleby-in-Westmorland, home of Trans-Pennine Publishing, was one of the fire stations that operated this Bedford C type water tender (FEC 78) bodied by HCB and supplied to the Westmorland County Fire Brigade in 1958. Westmorland was one of the shire counties that completely vanished under local government reorganisation in 1974. Vauxhall Motors

In conclusion we should note how the fire services had changed in the past 20 years, but even greater changes were to come. Up until this time it will be clear to the readers that virtually all the fire engine chassis makes were British, and it therefore came as some surprise when the Glasgow Fire Service broke the mould. Between 1967 and 1970 they purchased seven German Magirus Deutz chassis. Four were bodied by Scottish Motor Transport (SMT) as water tender escapes, whilst Bennetts (2) and Fulton & Wylie (1) provided the bodies on which three Magirus DLK30 turntable ladders were housed. This was the first time that a British fire brigade had looked abroad for their chassis but, within 20 years, it would be the Continental manufacturers that would hold the lion's share of the fire engine chassis market. How and why the British manufacturers fell by the wayside is another story which is still to be told.

During the course of writing this book and recording the history of the fire services during the 1950s and '60s I have received assistance from many quarters but the sources of information and photographs which have been the most influential are Simon Adamson and John Shakespeare.

It is true to say that, without their help and encouragement along with their vast fire service knowledge, it would not have been possible for me to write this book. Both have provided countless photographs from their collections and I look upon them almost as co-authors. We also owe much to the photographic record of the late-John Hughes, whose period pictures have added much to this publication. Other people have also willing given their time and expertise in providing all kinds of details to ensure the accuracy of this work and the following are sincerely acknowledged:

Andy Anderson, Aidan Fisher with the HCB-Angus archives, Mike Hebard, Barrie Lowe, Gary Chapman, Bill Aldridge and Vauxhall Motors, plus the many people whose names are credited under their photographs. There are countless others, too many to name individually, who have provided snippets and their help is also greatly appreciated. It should be stressed that, despite all the knowledge which is passed on in good faith, discrepancies do creep in and if any errors are included in the previous pages I apologise. However I would be more than grateful to learn of any untruths so that my own, and others, historical knowledge can be corrected.